Park Benjamin

Wrinkles and Recipes

Vol. 2

Park Benjamin

Wrinkles and Recipes
Vol. 2

ISBN/EAN: 9783337420246

Printed in Europe, USA, Canada, Australia, Japan

Cover: Foto ©berggeist007 / pixelio.de

More available books at **www.hansebooks.com**

WRINKLES

AND

RECIPES,

COMPILED FROM THE

SCIENTIFIC AMERICAN.

A Collection of Practical Suggestions, Processes, and Directions for

THE MECHANIC,

THE ENGINEER,

THE FARMER, AND

THE HOUSEKEEPER.

ILLUSTRATED.

EDITED BY

PARK BENJAMIN.

Revisers and Contributors:

PROFESSOR R. H. THURSTON, C.E.

PROFESSOR P. H. VANDER WEYDE, M.D.

RICHARD H. BUEL, *Mechanical Engineer.*

JOSHUA ROSE, *Mechanical Engineer.*

NEW-YORK:

H. N. MUNN, PUBLISHER,

37 Park Row.

If any worker in the great army of those in whose intelligent labor lies the surest foundation for the prosperity of us all, finds among the suggestions here compiled one thought, as it is hoped he may many, which shall lighten his toil or aid him in the production of better work, then the editor sincerely trusts that that worker will deem this little volume as prepared for his especial benefit, and to him especially dedicated.

PREFACE.

THE aim in the following pages has been to compile a collection of suggestions for the every-day use of the working-man, in his shop, about his dwelling, or in his household. The book is not an encyclopædia of recipes, nor does it make any pretensions to that title ; on the contrary, a very large number of formulæ have purposely been omitted, because they are, for the most part, attainable in other and more extensive works. Preference has been given to practical hints, and, while the majority of these have been carefully gathered and condensed from the back files of the *Scientific American*, and more especially from the letters of correspondents of that paper, a goodly proportion are entirely new and fresh, and have been prepared expressly for this book.

For the contribution of a large part of the matter in, and for the general supervision of the department of Mechanics, the editor is indebted to that thorough workman, Mr. Joshua Rose. To Mr. Richard H. Buel (whose articles are signed " B.") similar acknowledgments are owing for valuable papers on boilers, engines, and other topics in the department of Engineering. The general revision of the last-mentioned department has been the labor of Professor R. H. Thurston, of the Stevens Institute, as has the similar overlooking of the department of Technology, that of Professor P. H. Van der Weyde. To both of these distinguished gentlemen, as well as to Messrs. Munn & Co., the publishers of the *Scientific American*, who have most kindly afforded the facilities for the preparation of this work, the cordial acknowledgments of the editor are due.

NEW-YORK, Nov. 1, 1875.

CONTENTS.

MECHANICS.

MASTER-TOOLS: THEIR MANUFACTURE AND USE.

THE master-tools, here illustrated and described, comprise all that are necessary for plain machine-work in every description of metal ; and if they are made of the precise shape, and according to the given instructions, they will perform the full amount of duty here allotted to them, which, though it may appear to be unusually great, may be thoroughly relied upon for metal of any ordinary degree of hardness. Nor can any less amount of duty be obtained from them without evidencing inferior mechanical skill either in making or using the tool. It is true economy to obtain from a cutting tool its utmost amount of duty, which, though it may entail a little more drawing out and grinding of the tool in a given time, does not involve any more as compared to the quantity of work performed. It is well within the mark to say that at least one third more duty, in a given time, may be obtained from cutting-tools for metals (used in all machines having variable speeds and feeds, such as the lathe and the shaping-machine), than is obtained in the usual practice of our machine-shops, especially in the larger ones.

BORING-TOOL FOR BRASS.—The boring-tool for brass-work, here shown, is a standard tool for either roughing out or finish-

BORING-TOOL FOR BRASS.

ing, both of which duties it will perform equally well. It is bent further round at the end than is the boring-tool for wrought-iron, to prevent it from jarring or chattering. It is a master-tool in every sense of the word. It should be hardened right out, and used with a quick speed and light feed, no matter how deep the cut is. To prevent chattering or jarring when extending far out from the tool-post, or when it is very slight in body, it should have the top face depressed toward the cutting edge. When this tool is a stout one, the point may be ground more round, which will make it cut to finer finish.

BORING-TOOL FOR WROUGHT-IRON OR STEEL.—For turning out small holes, the tool here represented has no equal, providing it

BORING-TOOL FOR WROUGHT-IRON OR STEEL.

be made of the precise shape shown, the reasons for which are as follows : The cutting end must not be bent, in forging, any further round, because, in that case, the strain placed upon the tool by the cut will be in a direction tending to revolve the tool in the tool-post, giving the tool a corresponding tendency to spring away from its cut ; and further, because so stout a tool could not be got into the same size of hole. The degree of bend or angle of the centre-line of the bent end to the centre-line of the length of the body of the tool, causes the strain of the cut to be placed comparatively endwise of the tool, endeavoring to force it back into the tool-post, and thus places the strain in the direction in which the tool is best capable of withstanding it. The keenness and shape given to the top face of the tool make the cutting edge perform its duty on the front edge, which again tends to place the strain endwise on the tool, operating, by the strain on the top face of the tool (caused by its bending the shaving), to keep the tool to its cut by giving it an inclination to feed itself forward, thus relieving the feed-screw and nut of the slide-rest of a part of the duty of feeding. The cutting edge should not, even when the tool is newly forged, stand much, if any, above the horizontal plane of the top of the body of the tool, otherwise so stout a tool can not be got into a given size of hole, a consideration which is of the utmost importance : because boring-tools, from their comparative slightness, especially in long holes, are apt, under the most favorable of conditions, to spring away from the cut as the cutting proceeds toward the back end of the hole, thus making the latter a taper, of which the back end has the smallest diameter, necessitating several fine finishing cuts in order to make a parallel hole. If, however, every means is taken to use as stout a tool as the size of the hole will admit, the boring-tool will bore a very true and smooth hole.

In using these tools, it is best to employ a comparatively quick speed and light feed, no matter what the depth of the cut may be. They should be tempered to a very light straw if the tool is slight, and otherwise hardened right out; and the work should be freely supplied with soapy water. For use on copper, the top face should be ground more hollow, so that the cutting edge will be much more keen than is here shown. Whenever there is sufficient room in the hole, a stout bar of iron or steel should be held in the tool-post, and a short tool secured by set-screw in the end of the bar, thus securing greater rigidity than is possessed by a boring-tool, and facilitating the forging and grinding of the cutting-tool.

FINISHING-TOOL FOR CAST-IRON.—Cast-iron may be finished true and smoothly by a tool having a much broader cutting and scraping surface than is applicable to any other metal; and we are therefore enabled to apply to it, for finishing purposes, the

FINISHING-TOOL FOR CAST-IRON.

tool above illustrated, setting it so that its square nose is placed quite parallel with the work, and feeding it with a feed almost as coarse as the width of the square nose, say 8 revolutions of the lathe per in. of tool travel on small work, and 3 revolutions per ditto for large work. The tool is held with the cutting edge as close to the tool post as can possibly be convenient, and the cutting speed is about 25 to 30 feet per minute on small work, and 18 feet on large work, the tool being hardened right out in all cases.

FINISHING-TOOL FOR WROUGHT-IRON, CAST-IRON, OR STEEL.—This is a finishing-tool for wrought-iron which will cut smoothly, clean, and true. being far preferable to the square-nosed tools sometimes used for the purpose of finishing iron, since such tools do not turn wrought-iron true, but follow the texture of the metal, cutting deepest in the softer parts, especially when their edges become in the least dull from use. This tool should be held with the cutting edge as close in to the tool-post or clamp as it can conveniently be, with a quick speed and fine feed, soapy water being applied to the work. It may be also used for taking light roughing cuts on small work, and is, for such purposes, an excellent tool, especially upon work so slight as to be liable to spring, for which purpose the cutting point should not be much rounded. Ground very keen, it will answer admirably for copper work, the cutting speed being very great; that is to say, at least fourfold that given below, which is for finishing cuts on wrought or cast iron.

FINISHING-TOOL FOR WROUGHT-IRON, CAST-IRON, OR STEEL.

Size of work, inches diameter.	Cutting speed, feet per minute.	Feed.
1 and less	38	30
1 to 2	30	25
2 " 5	25	20
5 " 12	23	20
12 " 20	20	16
20 and over	18	14

This tool should always be hardened right out; and if used upon cast-iron, it should have less keenness upon the top face; that is to say, the plane of the top face should be ground more nearly to the same plane as the top face of the body of the tool. For use upon steel, the top face must be ground more nearly horizontal—a rule which, we may here observe, applies to all tools used upon wrought-iron. It should be placed in the lathe so that its cutting edge stands above the horizontal centre-line of the work.

FRONT-TOOL FOR BRASS-WORK.—This is a complete master-tool, filling every necessary qualification for all plain outside brass-work, and doing the duty on that metal which the front-tool and right and left hand side-tools do on wrought-iron. As shown in the engraving, it is ground to suit either roughing out r finishing. For very slight work, which is liable to spring, it may be ground a little more keen on the side faces, the top face not requiring, under any possible circumstances or conditions, to be ground keener than shown above. When held far out from the tool-post, the top face should be ground away, sloping down toward the cutting edge, which is done to prevent the tool from jarring or chattering. It should be hardened right out, and not

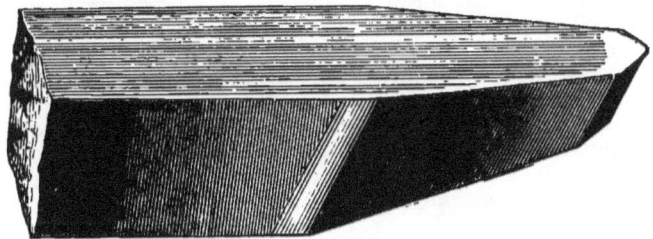

FRONT-TOOL FOR BRASS-WORK.

lowered or tempered at all, and used for roughing out at the following speed and feeds :

Size of work, in inches.	Cutting speed, in feet.	Feed.
1 and less	350	25
2 to 5	250	25
5 to 12	200	25
12 to 20	150	30

For finishing-cuts, the cutting speed may be increased by about one fifth, which rule will also apply to its use upon yellow brass for roughing out as well as finishing purposes.

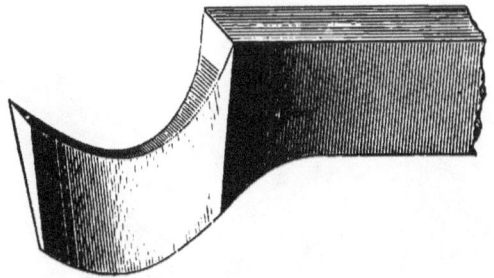

PARTING-TOOL.

PARTING OR GROOVING TOOL FOR IRON OR STEEL.—The parting tool is applicable either to cutting grooves or for parting, or, in other words, cutting work apart. The cutting point, or end of the tool, is made thicker than the metal, both vertically and horizontally, behind it, so that the latter shall clear and not grind against the sides of the groove. This tool, especially if made thin to suit some especial purpose, is excessively liable to spring, in consequence of the pressure of the cut ; and if it commences to spring, it is apt to dig into the cut, and then break from the excessive strain. It is to prevent this digging in that the top face of the cutting part of the tool is placed so much below the top face of the body of the tool, which may, however, be dispensed with when the cutting edge is held close in to the tool-post, and the grooving is not required to be very deep. When, however, these requirements do exist, the form illustrated is absolutely indispensable to rapid and reliable duty, whether the tool be used in a lathe or a planing-machine, the cutting edge of the tool being kept at about the horizontal centre of lathe-work, by packing-pieces placed beneath the body of the tool. If the width of the tool is not less than $\frac{3}{16}$ inch, and does not require to cut a groove deeper than $\frac{4}{8}$ inch, it should be hardened right out; if, however, these conditions are reversed, it should be tempered to a dark straw, and for very weak tools even to a purple color, as lowering the temper increases the strength of all tools. If the groove to be cut is sufficiently broad to cause the tool to spring, it is best to use a narrower one and cut it out in two separate cuts, moving the tool.

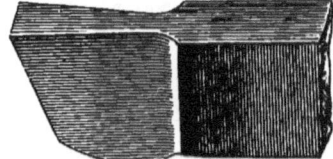

PARTING-TOOL FOR BRASS.

PARTING-TOOL FOR BRASS.—The parting-tool for brass is governed by the same principle as that for iron, save that its top face must be ground level, except in cases where the cutting edge stands far out from the tool-post, in which event the top face

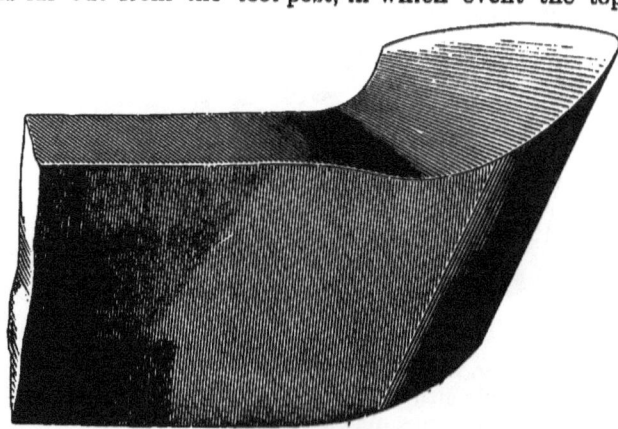

ROUGHING-TOOL FOR WROUGHT-IRON.

must be ground away at an angle of which the cutting edge is the lowest part. It is rarely, however, necessary for brass-work to grind the cutting edge much below the level of the top face of the body of the tool, as is shown for use on wrought-iron. The degree of hardness of the tool should be the same for brass as that given for wrought-iron.

ROUGHING-OUT FRONT TOOL FOR WROUGHT-IRON.—The engraving represents the best possible form of tool for roughing out wrought-iron, or for removing a large mass of that metal in the lathe or planing-machine. When used on large work, it should be tempered to a light straw-color, which will leave it strong enough to stand without breaking the heavy strain due to the cut. It must be held very firmly, and with the cutting edge as close to the tool-post as it can well be.

The following are its rates of cutting speed and feed, the speed meaning the length of shaving it cuts off, and the feed implying the number of revolutions of the lathe necessary to feed the tool an inch along its cut:

Size of work, inches diameter.	Cutting speed, feet per minute.	Feed.
1 and less	35	25
1 to 2	25	20
2 " 5	20	20
5 " 12	18	15
12 " 20	16	12
20 and over	15	12

For work of five inches diameter, and for all sizes below that, the tool should be hardened right out; that is, made as hard as fire and water will make it, and not tempered at all. For work of a larger size, it should be tempered to a light straw-color. This tool, with the top face ground less keen, that is, more nearly horizontal, is an excellent one for steel, and the harder the metal to be cut, the more nearly horizontal the top face must be. It should be placed, for lathe-work, so that the cutting edge stands a little above the horizontal centre-line of the work.

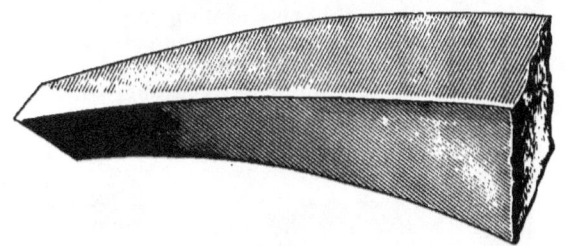

SIDE-TOOL FOR BRASS.

SIDE-TOOL FOR BRASS.—This tool fills the same place with reference to that metal that the side-tool and knife-tool do to iron-work; and it has no superior for taking out corners, for cutting out holes or recesses which do not pass entirely through the metal. In conjunction with the front-tool for brass, already illustrated, it will perform almost any duty upon either inside or outside brass-work, except, of course, cutting out narrow grooves. Its cross-section is somewhat diamond-shaped, and it is made right and left by bending in opposite directions. It is a far better tool than those bent round at the end after the manner of a boring-tool; and being more rigid, it is easier to forge and grind, and less liable to jar or chatter. It is equally applicable as a roughing-out or a finishing tool. It should be hardened right out, and used at the speeds and feeds given for the front-tool for brass.

A

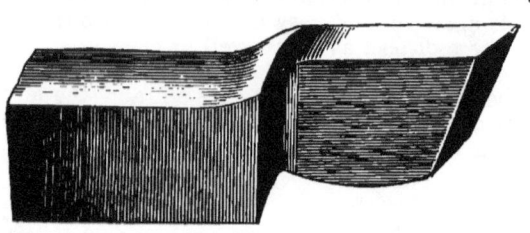

SIDE-TOOL FOR SQUARING ENDS OF WROUGHT-IRON WORK.

SIDE-TOOL FOR SQUARING ENDS OF WROUGHT-IRON WORK.—The illustration represents a side-tool for wrought-iron, to be employed for squaring the ends of work held between the lathe centres, and in other cases wherein there is not sufficient room to admit a stouter tool. The cutting edge is shown at A, and should be made more keen for wrought than for cast iron. In forging it, the hammering edgewise should be performed first, nor should any hammering be done to it edgewise after the steel has lost its redness. It should, for light duty, and for all finishing purposes, be hardened right out, and, for heavy duty, be tempered to a straw-color. If, however, this tool is employed, as it sometimes is, for very heavy duty on a slotting-machine, taking a cut, say, $2\frac{1}{2}$ inches deep and $\frac{1}{8}$ inch thick, it must be lowered to a brownish purple and used at a cutting speed of about 10 feet per minute, and be ground so that the cutting edge first strikes the cut near the body of the tool, and not at the point end. For ordinary work, it is best used with a comparatively fine feed and quick speed, since it is not sufficiently strong, when made very hard, to stand heavy duty.

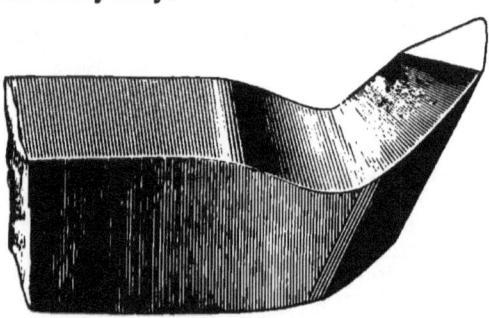

SIDE-TOOL FOR IRON.

SIDE-TOOL FOR WROUGHT-IRON, CAST-IRON, OR STEEL.—The engraving represents the most superior side-tool for either wrought or cast iron or steel, the only difference being that it requires to be less keen for the latter than, as here shown, for the former. It is employed to cut side-faces and to take out round or square corners. For small work, it should be made so that it will cut at the point, and not on both edges at one time, when used in a square corner. For heavy work, it may be made more round-nosed, and allowed to cut all round the curve; and it will, in either shape, work equally well as a roughing-out or as a finishing tool, only requiring to be ground more keen to fit it for finishing purposes (which should be, on wrought-iron, performed with soapy water applied to the work), and at a faster speed and finer feed. For taking out a round corner or fillet in slight work, which is liable to spring from the pressure due to the cut, the point must be rounded very little, the curve being made by operating both the straight and cross feeds of the lathe. This tool is made right or left handed by simply bending it in the required direction, that illustrated being a left-hand tool. It should be made as hard as fire and water will make it, and used at the following speeds and feeds:

Diameter of work, in inches.	Speed, in feet, per minute.	Feed.
1 and less	30	30
1 to 2	25	25
2 " 5	22	20
5 " 12	20	20

When, however, it is employed for roughing-out purposes, these speeds may be, with advantage, slightly diminished and the feeds increased.

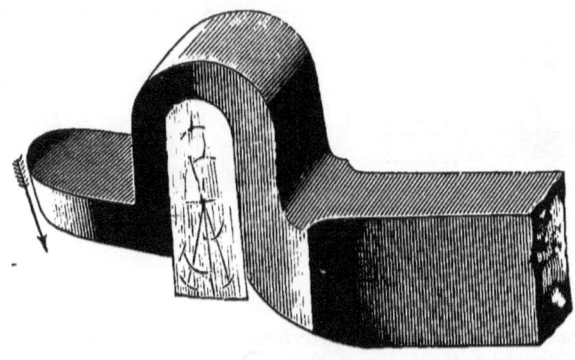

SPRING-TOOL.

SPRING-TOOL FOR USE ON WROUGHT-IRON, CAST-IRON, STEEL, OR BRASS.—The spring-tool is especially adapted to finishing sweeps, curves, or round or other corners, and will answer for any metal whatever. As illustrated, the face is given an upward incline to make it sufficiently keen for wrought-iron or steel. For brass-work, this face should be made horizontally level, or, if the cutting stands with its cutting edge far out from the tool-post, it may be inclined downward to make it cut smoothly. The piece of wood shown driven in the bend is to correct any tendency of the tool to spring away from the hard parts of the metal, as it is apt to do. The spring-tool does not turn so true as is desirable, still the smoothness of its finish makes it the most desirable tool for the purposes mentioned. It should have its face filed up very smoothly before being hardened, and should not be ground on that face. The application of the oilstone greatly improves its value for finishing. It should, for all purpose, be tempered to a brown color on the face, and left soft around the bend.

TOOLS.

BORING-BAR FOR LATHE-WORK.—In boring work chucked and revolved in the lathe, such, for instance, as axle-boxes for locomotives, the boring-bar here shown is an excellent tool. A represents a cutter-head, which slides along, at a close working fit, upon the bar, D D, and is provided with the cutters, B B B,

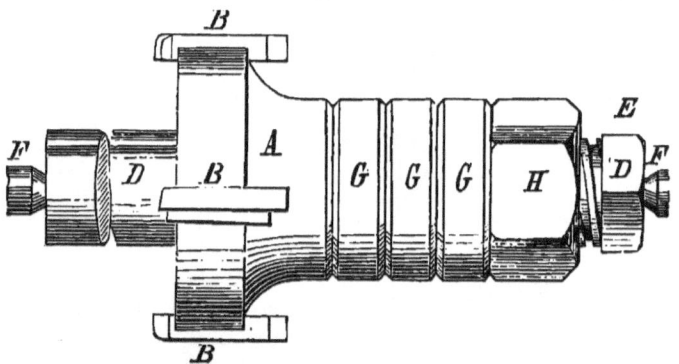

BORING-BAR.

which are fastened into slots provided in the head, A, by the keys shown. The bar, D D, has a thread cut upon part of its length, the remainder being plain, to fit the sliding head. One end is squared to receive a wrench, which, resting against the bed of the lathe, prevents the bar from revolving upon the lathe-centre, F F, by which the bar is held in the lathe. G G G are plain washers, provided to make up the distance between the thread and plain part of the bar, in cases where the sliding head, A, requires considerable lateral movement, there being more or fewer washers employed according to the distance along which the sliding head is required to move. The edges of these washers are chamfered off to prevent them from burring easily. To feed the cutters, the nut, H, is screwed up with a wrench.

The cutter-head, A, is provided in its bore with two feathers, which slide in grooves provided in the bar, D D, thus preventing the head from revolving upon the bar. It is obvious that this bar will, in consequence of its rigidity, take out a much heavier cut than would be possible with any boring-tool, and furthermore that, there being four cutters, they can be fed up four times as fast as would be possible with a single tool or cutter.

BORING-TOOLS, Shapes of.—The pressure on the cutting edge of a tool acts in two directions, the one vertical, the other lateral. The downward pressure remains at all times the same; the lateral pressure varies according to the direction of the plane of the cutting edge of the tool to the line or direction in which the tool travels: the general direction of the pressure being at a right angle to the general direction of the plane of the cutting edge. For example, the lateral pressure, and hence the spring of the various tools, shown in the cut, will be in each case in the direction denoted by the dotted lines. D is a section of a piece of metal requiring the three inside collars to be cut out;

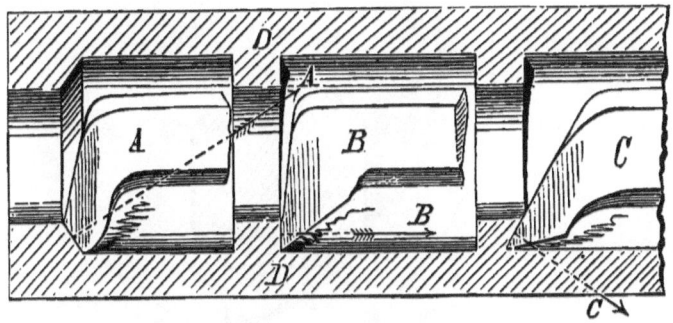

LATERAL PRESSURE OF TOOLS.

A, B, and C are variously shaped boring-tools, from which it will be seen that A would leave the cut in proportion as it suffered from spring, which would increase as the tool edge became dull, and that the cut forms a wedge, tending to force the tool toward the centre of the work. B would neither spring into nor away from the cut, but would simply require more power to feed it as the edge became dulled; while C would have a tendency to run into the cut in proportion as it springs; and as the tool edge became dull, it would force the tool-point deeper and deeper into the cut until something gave way. Now, in addition to this consideration of spring, we have the relative keenness of the tools, it being obvious at a glance that (independent of any top rake or lip) C is the keenest and A the least keen tool; and since wrought-iron requires the keenest, cast-iron a medium, and brass the least keen tool, it follows that we may accept, as a rule, C for wrought iron, B for cast-iron, and A for brass-work. In Fig. 2, B represents a section of the work, No. 1 represents a boring-tool with top-rake, for wrought-iron, and No. 2 a tool without top-rake, for brass-work, which may be also used for cast-iron when the tool stands a long way out from the tool post or clamp, under which circumstances it is liable to jar or chatter. A tool for use on wrought-iron should have the same amount of top-rake, no matter how far it stands out from the tool-post; whereas one for use on cast-iron or brass requires to be the less keen the further it stands out from the tool-post.   To take a very smooth cut on brass-work, the top face of the tool, shown at 2 in Fig. 2, must be ground off, as denoted by the dotted line.

We have now to consider the most desirable shape for the corner of the cutting edge.   A positively sharp corner, unless for

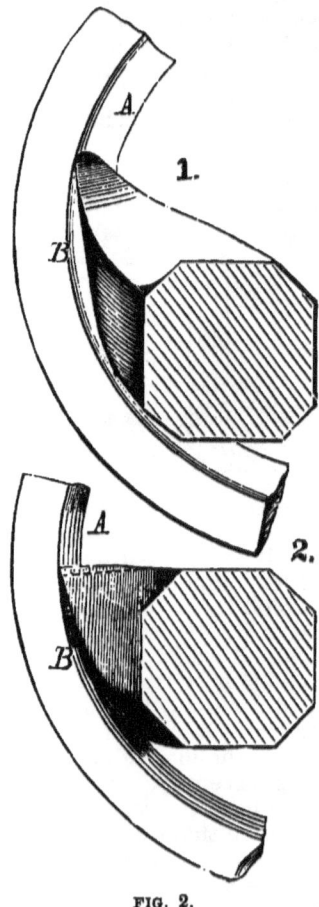

FIG. 2.

a special purpose, is very undesirable, because the extreme point soon wears away, leaving the cutting qualification of the tool almost destroyed, and because it leaves the work rough, and can only be employed with a very fine feed.   It may be accepted as a general rule that, for roughing cuts, the corner should be sufficiently rounded to give strength to the tool-point; while, in finishing cuts, the point may be made as round as possible without causing the tool to jar or chatter.   Now, since the tendency of the tool to jar or chatter depends upon four points, namely, the distance it stands out from the tool-post, the amount of top-rake, the acuteness or keenness of the general outline of the tool, and the shape of the cutting corner, it will be readily seen that

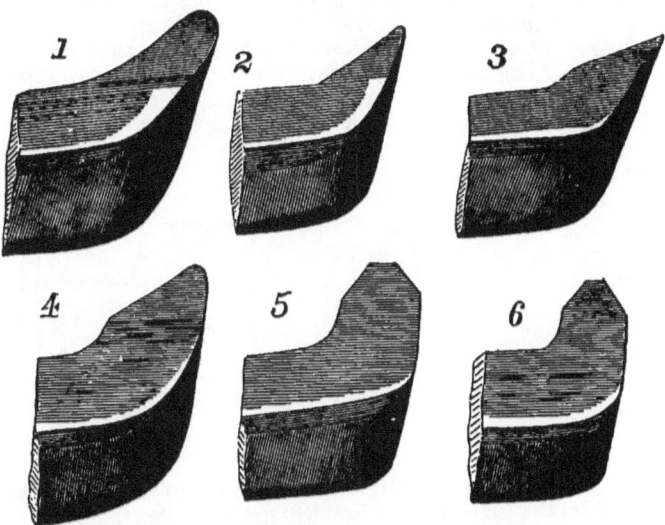

FIG. 3.—VARIOUS FORMS OF BORING-TOOLS.

judgment is required to determine the most desirable form for any particular conditions, and that it is only by understanding the principles governing the conditions that a tool to suit them may be at once formed. In Fig. 3 will be found the various forms of boring-tools for ordinary use. No. 1 is for use when the conditions admit of a heavy cut on wrought-iron. No. 2 is for use on wrought-iron when the tool stands so far from the tool-post as to be necessarily subject to spring. No. 3 is to cut out a square corner at the bottom of a hole in wrought-iron. No. 4 is for taking out a heavy cut in cast-iron. No. 5 is for taking out a finishing cut in cast-iron when the tool is proportionally stout, and hence not liable to spring or chatter; the point being flat, the cutting being performed by the front corner, and the back part being adjusted to merely scrape. No. 6 is for use on cast-iron under conditions in which the tool is liable to jar or spring.

CHILLED CAST-IRON TOOLS, To cast, for cutting chilled iron.— Make a tool of the required shape out of wrought iron, then cast the chilled part, using charcoal-iron No. 5.

CHIPPING.—The chisel requires special notice since it is frequently made of the most ill-advised shape (for either cutting smoothly or standing the effects of the blow), that is, hollow, as in Fig. 1, in which case there are two sections of metal, represented by the dotted lines, *a a*, which are very liable to break, from their weakness and from the strain outward placed upon them by the cut, which, acting as a wedge, endeavors at each blow to drive them outward instead of inward, as would be the case in a properly shaped chisel, as shown in Fig. 2, *a* being the cutting edge.

When using, hold it firmly against the cut, and it will do its work smoother and quicker.

The cape, or, as it is sometimes called, cross-cut chisel, is employed to cut furrows across the work to be chipped, which furrows, being cut at a distance from each other less in width than the breadth of the flat chisel, relieve the flat chisel and prevent its corners from "digging in" and breaking. If a large body of metal requires to be chipped off cast-iron or brass, the use of the

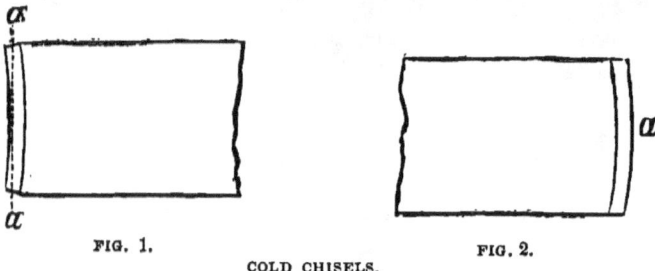

FIG. 1.

COLD CHISELS.

FIG. 2.

cape-chisel becomes especially advantageous, for the metal, being weakened by the furrows, will break away in pieces from the force of the blow, without requiring to be positively cut by the chisel ; but care must be taken to leave sufficient metal to take a clean finishing cut ; for when the metal is broken away, by the force of the blow, it is apt to break out below the level of the cut. It is also necessary to nick deeply with a chisel the outside edges of the work at the line representing the depth of the metal to be chipped off, so that the metal shall not break away at the edges deeper than the cut is intended to be.

CHISELS, COLD, To grind.—Grind a cold-chisel so that its cutting edge is rounding and not hollow, as it is often made. A rounded chisel is stronger and cuts smoother. A hollow chisel has no useful place as a chipping-tool.

CHISELS, COLD, To harden.—Heat the chisel to a distance about equal to its width, immerse it about half that distance in the water, hold it still about four seconds, suddenly dip it a little deeper, and then withdraw. Brighten one flat surface with a piece of grindstone or an emery-block ; then brush the hand or a piece of waste over the brightened surface to remove the false color, and finally cool out in the water, when the clear blue color appears.

CHISELS, COLD, To hold.—Hold a cold-chisel firmly to its cut, without removing it at every blow. This will increase its effectiveness, and decrease its liability to break from a foul blow.

CHISELS, COLD, To remove burrs from the heads of.—When the head of a cold-chisel is burred over from use, rest the head upon a block of iron, and strike the burrs from the under side, and they will break readily and easily off.

CHISELS, COLD, Use of.—These should be kept thin at the cutting end, which saves time and entails less labor in using.

CONE-PLATE for boring in the lathe.—For chucking shafts and other similar work in the lathe (to bore holes in the ends of the shafts, etc.), the cone-plate shown in the engraving is the best appliance known to machinists. A is a standard, fitting in the shears of the lathe, at E, and holding the circular plate, C, by means of the bolt, B, which should be made to just clamp the plate, C, tightly when the nut is screwed tight. The plate contains a series of conical holes, 1, 2, 3, etc. (shown in section at D). The object of coning the pin, B, where it carries the plate, C, is that the latter shall be made to a good working fit and have no play. The operation is to place the shaft in the lathe, one end being provided with a driver, dog, or carrier, and placed on the running or line centre of the lathe; and the other end, to be

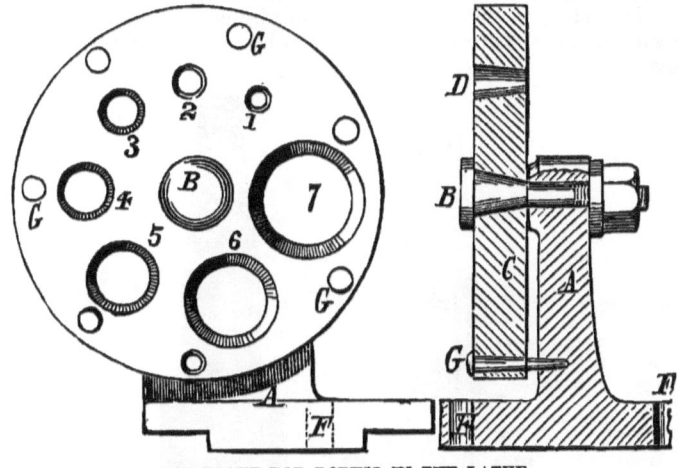

CONE-PLATE FOR BORING IN THE LATHE.

operated upon, being placed in such one of the conical holes of the plate, C, as is of suitable size, the distance of the standard, A, from the lathe-centre is to be adjusted so that the work will revolve in the coned hole with about as much friction as it would have were it placed between both the lathe-centres. Thus the conical hole will take the place of the dead-centre of the lathe, leaving the end of the shaft free to be operated on. F F are holes to bolt the standard, A, to the lathe shears or bed; and G G, etc., are taper-holes to receive the pin, G, shown in the sectional view. The object of these holes and pin is to adjust the conical holes so that they will stand dead true with the lathe-centres; for if they stood otherwise, the holes would not be bored straight in the work. In the engraving, hole No. 7 is shown in position to operate, the pin, G, locking the plate, C, in that position. In setting the work, the nut on the pin, B, should be eased back just sufficiently to allow the plate, C, to revolve by hand; the work should then be put into position, and the pin, G, put into place; the standard, A, should then be adjusted to its distance from the live lathe-centre, and bolted to the lathe-bed ; and finally, the nut on the pin, B, should be screwed up tight, when the work will be held true, and the cone-plate prevented from springing. Care must be taken to supply the conical holes, in which the work revolves, with a liberal quantity of oil, otherwise they will be apt to abrade.

CUTTERS AND REAMERS, To prevent, cracking when being hardened.—Drill a small hole from the side to meet the large one at its enclosed termination.

CUTTERS FOR BOILER-PLATE AND SIMILAR WORK.—Cutters are steel bits, usually held in either a stock or bar, being fitted and keyed to the same ; by this means, cutters of various shapes and sizes may be made to fit one stock or bar, thus obviating the necessity of having a multiplicity of these tools. Of cutter-stocks, which are usually employed to cut out holes of comparatively large diameter, as in the case of tube-plates for boilers, there are two kinds, the simplest and easiest to be made being that shown below.

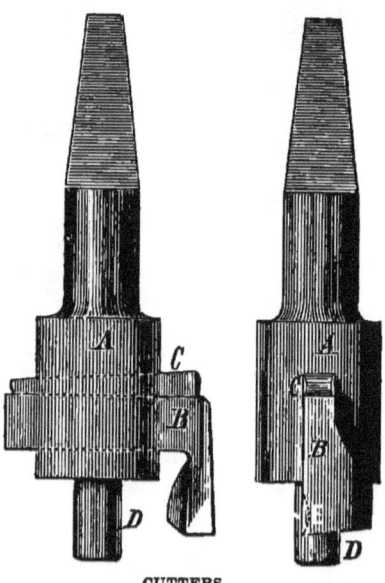

CUTTERS.

A is the stock, through which runs a slot or keyway into which the cutter, B, fits, being locked by the key, C. D is a pin to steady the tool while it is in operation. Holes of the size of the pin, D, are first drilled in the work, into which the pin fits. To obviate the necessity of drilling these holes, some modern drill-stocks have, in place of the pin, D, a conical-ended pin which acts as a centre, and which fits into a centre-punch mark made in the centre of the hole to be cut in the work. Most of these devices are patented, and the principle upon which they act will be understood from the second engraving, A being the stock to which the cutters, B B, are bolted with one or more screws. C is a spiral spring working in a hole in the stock to receive it. Into the outer end of this hole fits, at a working fit, the centre, D, which is prevented from being forced out (from the pressure of the spring,

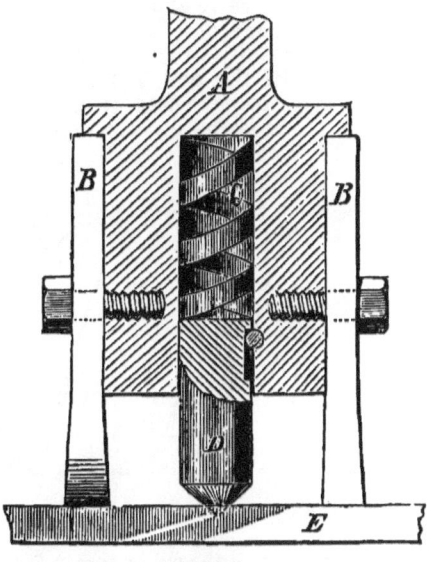

CUTTERS.

C) by the pin working in the recess, as shown. E is the plate to be cut out, from which it will be observed that the centre, D, is forced into the centre-punch mark in the plate by the spring, C, and thus serves as a guide to steady the cutters and cause them to revolve in a true circle, so that the necessity of first drilling a hole, as required in the employment of the form of stock shown in the first figure, is obviated.

CUTTING SQUARE THREADS, Tools for.—For cutting square threads, the tool here represented is used. The point at C is

Fig. 14.

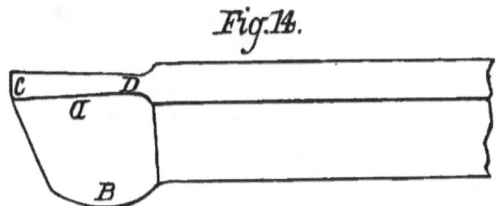

TOOL FOR CUTTING SQUARE THREADS.

made thicker than the width at D, so as to give the sides clearance from the sides of the thread. At B it is made thinner, to give the tool clearance, and deeper to compensate in some measure for the lack of substance in the thickness. The top face may, for wrought-iron or steel, be ground hollow, C being the highest point, to make it cut cleaner ; while, when held far out from the tool-post for use on brass-work, the face, C D, may be ground at an incline, of which C is the lowest point, which will prevent the tool from springing into the work.

If the pitch of the screw to be cut is very coarse, a tool nearly one half of the width of the space between one thread and the next should be employed, so as to avoid the spring which a tool of the full width would undergo. After taking several cuts, the tool must be moved laterally to the amount of its width, and cuts taken off as before until the tool has cut somewhat deeper than it did before being moved, when it must be placed back again into its first position, and the process repeated until the required depth of thread is attained.

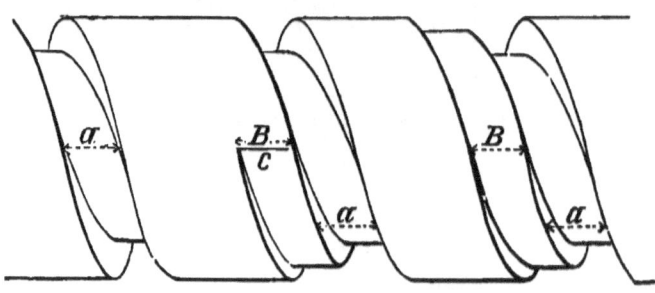

SCREW DURING CUTTING

The above figure represents a thread or screw during the above-described process of cutting. *a a a* is the groove or space taken out by the cuts before the tool was moved ; B B represents the first cut taken after it was moved ; *c* is the point to which the cut, B, is supposed (for the purpose of this illustration) to have traveled.

The tool used having been a little less than one half the proper width of the space of the thread, it becomes evident that the thread will be left with rather more than its proper thickness, which is done to allow finishing-cuts to be taken upon its sides, for which purpose the knife-tool already described is brought into requisition, care being taken that it is placed true, so as to cut both sides of the thread of an equal angle to the centre-line of the screw.

Adjustable dies, that is, those which take more than one cut to make a full thread, should never be used in cases where a solid die will answer the purpose, because adjustable dies take every cut at a different angle to the centre-line of the bolt, as explained by the following engravings.

The first represents an ordinary screw. It is evident that the pitch from *a* to B is the same as from C to D, the one being the top, the other the bottom, of the thread. It is also evident that a piece of cord wound once around the top of the thread will be longer than one wound once around the bottom of the thread, and yet, in passing once around the thread, the latter advanced as much forward as the former, that is, to the amount of the pitch of the thread. To illustrate this fact, let *a b*, in the following diagram, represent the centre-line of the bolt lengthwise, and

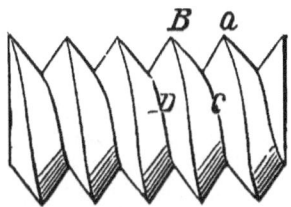

ORDINARY SCREW.

c d a line at right angles to it; then let from the point *e* to the point *f* represent the circumference of the top of the thread, and from *e* to *g* the circumference of the bottom of the thread, the lines *h h* representing their respective pitches; and we have the line *k* as representing the angle of the top of the thread to the centre-line, *a b*, of the bolt, and the line *l* as representing the angle of the bottom of the thread to the centre-line, *a b*, of the bolt, from which it becomes apparent that the top and the bottom of the thread are at different angles to the centre-line of the bolt. The tops of the teeth of adjustable dies are themselves at the greatest angle, while they commence to cut the thread on the bolt at its largest diameter, where it possesses the least angle, so

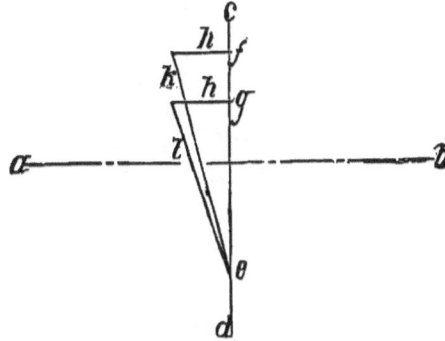

that the dies cut a wrong angle at first, and gradually approach the correct angle as they cut the depth of the thread. From what has been already said, it will be perceived that the angle of thread cut by the first cuts taken by adjustable dies, is neither that of the teeth of the dies nor that required by the bolt, so that the dies can not cut clean because the teeth do not fit the grooves they cut, and drag in consequence.

Dies for use in hand-stocks are cut from hubs of a larger diameter than the size of bolt the dies are intended to cut: this being done to cause the dies to cut at the cutting edges of the teeth which are at or near the centre of each die, so that the threads on each side of each die act as guides to steady the dies and prevent them from wabbling, as they otherwise would do; the result of this is that the angle of the thread in the dies is not the correct angle for the thread of the bolt, even when the dies are the closest together, and hence taking the finishing-cuts on the thread, although the dies are nearer the correct angle when in that position than in any other. A very little practice at cutting threads with stocks and dies will demonstrate that the tops of the threads on a bolt cut by them are larger than was the diameter of the bolt before the thread was commenced to be cut, which arises from the pressure placed on the sides of the thread of the bolt by the sides of the thread on the dies, in consequence of the difference in their angles; which pressure compresses the sides of the bolt-thread (the metal being softer than that of the dies), and causes a corresponding increase in its diameter. It is in consequence of the variation of angle in adjustable dies that a square thread can not be cut by them, and that they do not cut a good V-thread.

In the case of a solid die, the teeth or threads are cut by a hub the correct size, and they therefore stand at the proper angle; furthermore, each diameter in the depth of the teeth of the die cuts the corresponding diameter on the bolt, so that there is no strain upon the sides of the thread save that due to the force necessary to cut the metal of the bolt-thread.

CUTTING VERSUS SCRAPING TOOLS.—A tool will either cut or scrape, according to the position in which it is held, as, for instance, below the line A in the illustration. Line A is in each

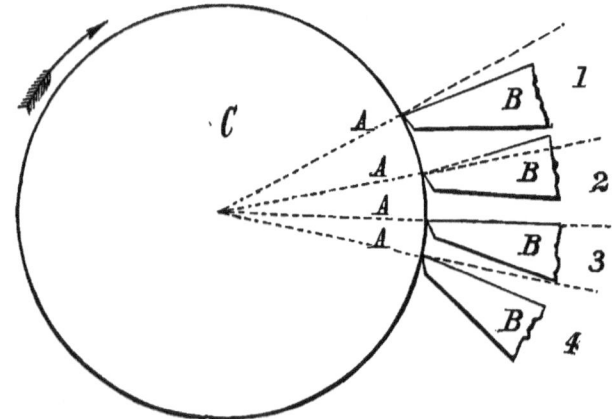

CUTTING AND SCRAPING.

case one drawn from the centre of the work to the point of contact between the tool edge and the work, C being the work, and

B the tool. It will be observed that the angle of the top face of the tool varies in each case with the line A. In position 1, the tool is a cutting one ; in 2, it is a scraper ; in 3, it is a tool which is a cutter and scraper combined, since it will actually perform both functions at one and the same time ; and in 4, it is a good cutting tool, the shapes and angles of the tools being the same in each case.

DIES, Fitting, to quadrants when either the dies or quadrants, or both, are to be hardened.—Make them a shade too small, to allow for their swelling during the process of hardening.

DIES, To ease hardened, that fit too tightly.—Supply them with very fine emery and oil, and work them backward and forward in their place along the travel.

DIES, To renew worn-out.—Slightly close the holes by swaging, fill the clearance-holes with Babbitt-metal, and recut them with the hub.

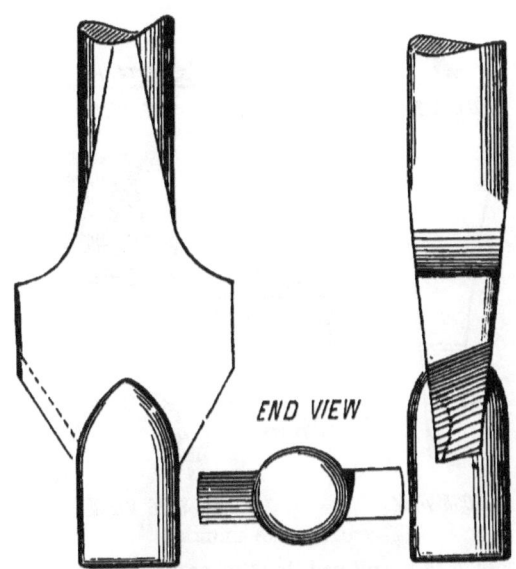

END VIEW

COUNTERSINK-PIN DRILLS.

DRILLS, COUNTERSINK-PIN.—Of these tools there are various forms. The following figure represents a taper countersink such as is employed for holes to receive flush rivets or countersunk head-bolts, this form of tool being mainly employed for holes above $\frac{5}{16}$ of an inch in diameter. It should be made, tempered, and used as directed for pin-drills. In tempering these tools, or any others having a pin or projection to serve as a guide in a hole, the tool should be hardened right out from the end of the pin to about $\frac{3}{8}$ of an inch above the cutting edges. Then lower the temper of the metal (most at and near the cutting-edges), leaving the pin of a light straw-color, which may be accomplished by pouring a little oil upon it during the lowering or tempering process. The object of this is to preserve it as much as possible from the wear due to its friction against the sides of the hole. For use on wrought-iron and steel, this countersink (as also the pin-drill) may have the front face hollowed out, after the fashion of the lip-drill, and as shown by the dotted line.

DRILLS, SLOT, for keyways.—For drilling out oblong holes, such as keyways, or for cutting out recesses such as are required to receive short feathers in shafts, the drill known as a slotting-drill, here shown, is brought into requisition. No. 1 is the form in

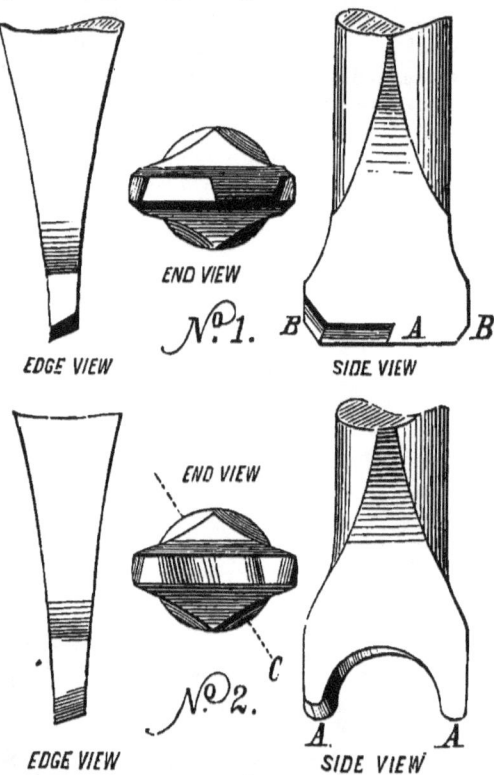

FIG. 1.—SLOT-DRILLS.

which this tool was employed in the early days of its introduction; it is the stronger form of the two, and will take the heaviest cut. The objection to it, however, is that, in cutting out deep slots, it is apt to drill out of true, the hole gradually running to one side. The method of using these drills is to move the work back and forth, in a chuck provided for the purpose, the drill remaining stationary. If these tools were used as common drills, they would cut holes of the form shown in Fig. 2.

FILES, The use of.—In draw-filing, take short, quick strokes, which will prevent the file from pinning and scratching. Long strokes, no matter how long the work may be, are useless save to make scratches. Remember that it is less the number of strokes you give the file than the weight you place upon it that is effective ; therefore, when using a rough file, stand sufficiently away from the vise to bring the weight of the body upon the forward stroke. New files should be used at first upon broad surfaces, since narrow edges are apt to break the teeth if they have the fibrous edges unworn. For brass-work, use the file on a broad surface

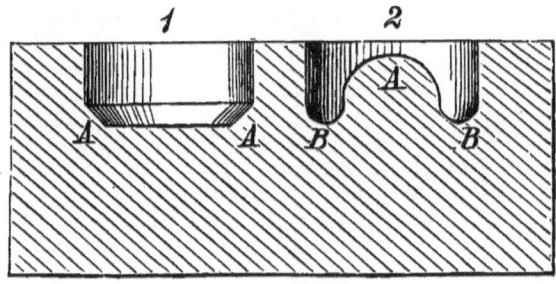

FIG. 2.—SLOT-DRILLS.

until ts teeth are dulled, then make two or three strokes of the file under a heavy pressure upon the edge of a piece of sheet-iron, which will break off the dulled edges of the teeth, and leave a new fibrous edge for the brass-work. Use bastard-cut files to take off a quantity of metal of ordinary hardness ; second-cut, in fitting, and also to file unusually hard metal ; smoothing, to finish in final adjustment or preparatory to applying emery-cloth ; dead smooth, to finish very fine work ; float-file on lathe-work.

FILES, To prevent scratching by.—To prevent files from pinning, and hence from scratching, properly clean them, and then chalk them well.

FILES, To resharpen.—(1) Carefully clean them with hot water and soda, then place them in connection with the positive pole of a battery in a bath composed of 4 parts of sulphuric acid and 100 parts water. The negative pole is to be formed of a copper spiral, surrounding the files but not touching them ; the coil terminates in a wire reaching above the surface. Leave the files in the bath ten minutes, then carefully wash them off, dry, and oil. (2) Carefully clean the files in hot water and soda, and dip for 40 minutes in nitric acid of 41° B.

FILES, To select.—To choose a flat file, turn it edge upward, and look along it, selecting those that have an even sweep from end to end, and having no flat places or hollows. To choose a half-round file, turn the edge upward, look along it, and select that which has an even sweep, and no flat or hollow places on the half-round side, even though it be hollow in the length of the flat side.

GRAVERS, Grinding.—Gravers should be ground on the end only, and not on the sides of the square, except when applied to brass-work.

HALF-ROUND BITS.—For drilling or boring holes very true and parallel in the lathe, the half-round bit shown in the engraving is·unsurpassed. The cutting edge, A, is made by backing off the end, as denoted by the space between the lower end of the tool and the dotted line, B, and performing its duty along the radius, as denoted by the dotted line in the end and top views. It is only necessary to start the half-round bit true, to insure its boring a hole of any depth true, parallel, and very smooth. To start it, the face of the work should, if circumstances permit, be made true ; this is not, however, positively necessary. A recess, true and of the same diameter as the bit, should be turned in the work, the bit then being placed in position, and the dead-centre employed to feed it to its duty, which (if the end of the bit is square, if a flat place be filed upon it, or any other method of holding it sufficiently tight be employed) may be made as heavy as the belt will drive. So simple, positive, and effective is the operation of this bit that (beyond starting it true and using it at a moderate cutting speed, with oil for wrought-iron and steel) no further instructions need be given for its use.

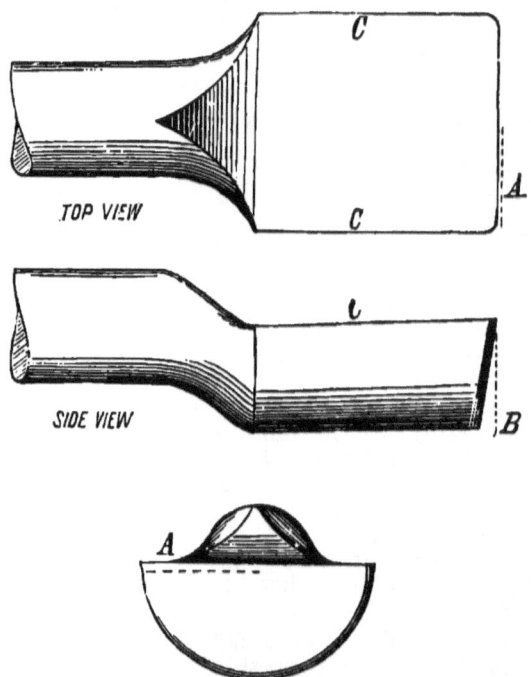

FIG. 1.—HALF-ROUND BITS.

In Fig. 2 is shown A, a boring-bar ; B B is the sliding head ; C C is the bore of the cylinder, and 1, 2, and 3 are tools in the positions shown. D D D are projections in the bore of the cylinder, causing an excessive amount of duty to be placed upon the cutters, as sometimes occurs when a cut of medium depth has been started. Such a cut increases on one side of the bore of the work until, becoming excessive, it causes the bar to tremble and the cutters to chatter. In such a case, tool and position No. 1 would not be relieved of any duty, though it spring to a considerable degree ; because the bar would spring in the direction denoted by the dotted line and arrow E, while the spring of the tool itself would be in the direction of the arrow, F. The tendency of the spring of the bar is to force the tool deeper into the cut instead of relieving it ; while the tendency of the spring of the tool will scarcely affect the depth of the cut. Tool and position No. 2 would cause the bar to spring in the direction of the dotted line and arrow G, and the tool itself to spring in the direction of H, the spring of the bar being in a direction to increase. and that of the tool to diminish, the cut. Tool and position No. 3 would, however, place the spring of the bar in a direc-

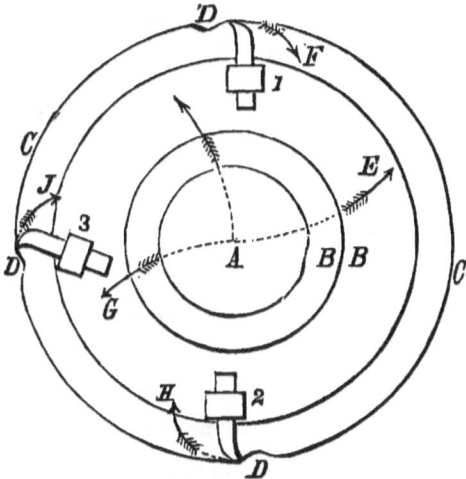

FIG. 2.—HALF-ROUND BITS.

tion which would scarcely affect the depth of the cut, while the spring of the tool itself would be in a direction to give decided relief by springing away from its excessive duty. It must be borne in mind that even a stout bar of medium length will spring considerably from an ordinary roughing-out cut, though the latter be of an equal depth all round the bore, and from end to end of the work. Position No. 3, in Fig. 2, then, is decidedly preferable for the roughing-out cuts. In the finishing cuts, which should be very light ones, neither the bar nor the tool is so much affected by springing ; but even here position No. 3 maintains its superiority, because, the tool being pulled, it operates somewhat as a scraper (though it may be as keen in shape as the other tools), and hence it cuts more smoothly. It possesses, it is true, the defect that the distance from the cutting point stands further out from the holding-clamp, and the tool is hence more apt to spring ; and in cases where the diameter of the sliding head is much less than that of the hole to be bored, this defect may possess importance, and then position No. 2 may be prefer-

able ; but it is an error to employ a bar of small diameter compared to that of the work.

To obtain the very best and most rapid result, there should be
but little space between the sliding head and the bore of the
work ; the bar itself should be as stout as is practicable, leaving
the sliding head of sufficient strength ; and if the bar revolves in
journals, these should be of large diameter and with ample facilities for taking up both the diametrical and end play of the boxes,
since the one steadies the bar while it is performing boring duty,
and the other while it is facing off end faces, as for cylinder-cover
joints.

HAMMERS, CHIPPING.—These should weigh about 1¾ lbs., and
have handles 15 inches long.

METAL HAVING A VERY HARD SKIN, Tools used for.—When
the skin of the metal to be cut is unusually hard, as frequently occurs in cast-iron, the shape of the cutting part of the boring-tool
must be such that its point will enter the cut first, so that it cuts
the inside and softer metal. The hard outside metal will then
break off with the shaving without requiring to be cut by the tool
edge, while the angle of the cut will keep the tool point into its
cut from the pressure required to break the shaving. A tool of

FIG. 1.—TOOL FOR HARD METAL.

this description is represented in the engraving, Fig. 1. a is the
point of the tool, and from a to B is the cutting edge ; the dotted
lines, c and D, represent the depth of the cut, c being the inside
skin of the metal, supposed to be hard.   The angle at which the
cutting edge stands to the cut causes the pressure, due to the
bending and fracturing of the shaving, to be in the direction of e,
which keeps the tool point into its cut ; while the resistance of
the tool point to this force, reacting upon the cut, from a to B,
causes the hard skin to break away.   For use on wrought-iron,
however, the tool presented below will work to better advantage,
it being a side-tool.   In the event of a side face being very hard,
it possesses the advantage that the point of the tool may be made
to enter the cut first, and, cutting beneath the hard skin, fracture
it off without cutting it, the pressure of the shaving on the tool
keeping the latter to its cut, as shown in Fig. 2.

a is the cutting part of the tool ; B is a shaft with a collar on
it ; c is the side cut being taken off the collar, and D is the face,
supposed to be hard.   The cut is here shown as being commenced from the largest diameter of the collar, and being fed inward
so that the point of the tool may cut well beneath the hard face,
D, and so that the pressure of the cut on the tool may keep it to

its cut, as already explained ; but the tool will cut equally as advantageously if the cut is commenced at the smallest diameter of the collar and fed outward, if the skin, D, is not unusually hard.

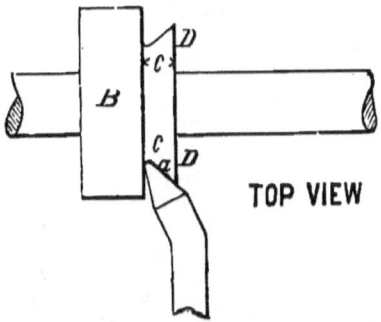

TOP VIEW

FIG. 2.—TOOL FOR HARD METAL.

PLANE-IRONS, Carpenters', To grind.—The angle of a plane-iron should be about 25°.

PLANE-IRONS, Angle of, to face of planes.—For soft wood, 50° ; for mahogany, 50° ; for soft wood for mouldings, 55° ; for hard

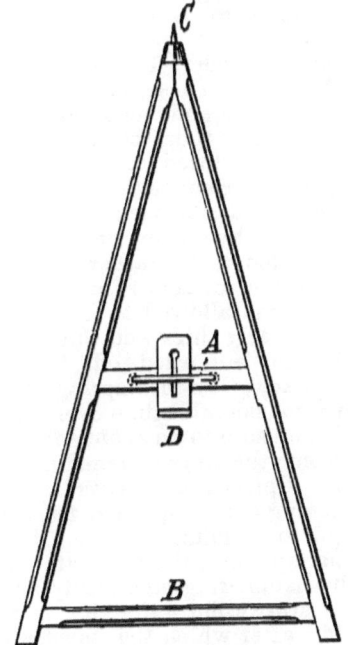

DEVICE FOR GRINDING PLANE-IRONS.

wood for mouldings, 60°. With this device, one man can both turn the stone and grind the tool much more accurately than by holding it in his hand.

A is a piece of spring-steel, 8 inches long, bent at each end, with thumbscrew. You grasp the holder with the left hand at B, sticking the point, C, into a board or the wall, at such a distance from the stone as to bring the iron, D, in the right position on the stone. By raising or lowering C, the bevel is regulated.

SAWS AND SPRINGS, Hardening.—The usual way of proceeding is to heat the saws in long furnaces, and then to immerse them horizontally and edgewise in a long trough containing the composition; two troughs are commonly used alternately. Part of the composition is wiped off with a piece of leather, when the articles are removed from the trough; they are then heated one by one over a clear coke fire until the grease inflames; this is called "blazing off." A greatly recommended composition consists of 2 lbs. of suet and ¼ lb. of beeswax to every gallon of whale oil; these are boiled together, and will serve for thin articles and most kinds of steel. The addition of black resin, to the extent of about 1 lb. to the gallon, makes it serve for thicker pieces, and for those it refused to harden before; but the resin should be added with judgment, or the articles will become too hard and brittle. The composition is useless when it has been constantly employed for about a month; the period depends, however, on the extent to which it is used, and the trough should be thoroughly cleaned out before the new mixture is placed in it. The following recipe is commended: 20 galls. spermaceti-oil, 20 lbs. melted and strained beef suet, 1 gall. neats'-foot oil, 1 lb. pitch, 3 lbs. black resin. These last two articles must be previously melted together, and then added to the other ingredients; the whole must then be heated in a proper iron vessel, with a close cover fitted to it, until the moisture is entirely evaporated, and the composition will take fire on a flaming body being presented to its surface; the flame must be instantly extinguished again by putting on the cover of the vessel. When the saws are wanted to be rather hard, but little of the grease is burned off; when milder, a large portion; and for a spring temper, the whole is allowed to burn away. When the work is thick, or irregularly thick and thin, as in some springs, a second and third dose is burned off, to insure equality of temper at all points alike. Gun-lock springs are sometimes literally fried in oil for a considerable time over a fire in an iron tray. The thick parts are then sure to be sufficiently reduced, and the thin parts do not become the more softened from the continuance of the blazing heat. Springs and saws appear to lose their elasticity after hardening and tempering, from the reduction and friction they undergo in grinding and polishing. Toward the conclusion of the manufacture, the elasticity of the saw is restored principally by hammering, and partly by heating over a clear coke fire, to a straw-color; the tint is removed by very diluted muriatic acid; after which the saws are well washed in plain water and dried.

SAWS, BAND, To solder.—Scarf the end of the saw to a taper for a distance of two fine or one coarse tooth, fitting the edges of the scarf very true and level. Clean the joint with acid, and clamp the saw ends together with a suitable frame or clamp. Heat the joint with a pair of red-hot tongs, and place a small strip of jeweler's silver solder upon the joint; press it upon the same with the red-hot tongs. As soon as the solder has properly run or flowed, remove the tongs and cool the joint with water, to restore as far as possible the temper of the saw. Then file the joint to an even thickness and level with the rest of the saw, using a wire gauge as a template.

SAWS, CIRCULAR, Speeds for :

Diameter of saw, inches.	Revolutions per minute for English or thin saws.	Revolutions per minute for American or thick saws.
36	1000	1500
30	1200	1800·
25	1400	2100
20	1500	2400
15	1800	2700
10	2900	3000

Circular saws of over 40 or 50 inches in diameter are, or should be, hammered to run at a certain speed. This is more important when the speed is as high as from 700 or 900 revolutions per minute. If a saw is so hammered as to do good work at 300 or 400 revolutions per minute, it will not do as good work at 900, for the reason that the high speed expands the outside or rim, causing it to dish, or "flop around," as sawyers sometimes express it. In such cases, and when it is inconvenient to reduce the speed, it will be necessary to guide the saw out of the log so as to cause the central part to rub against the log enough to heat it slightly, thus expanding the portion that needs hammering. An expert sawyer can in this way manage indifferently well, though at an expense of considerably more power. A large saw, to run well at high speed, should be hammered in the centre part until it is slightly dishing, or, as it is variously expressed, "loose at the eye," or "rim-bound." It may be loose at the eye when it is the reverse of rim-bound, namely, too open at the rim, which is the most frequent trouble with such saws. They all become so eventually from use, and should then be rehammered.

SAWS, Hints concerning.—A saw just large enough to cut through a board will require less power than a saw larger, the number of teeth, speed, and thickness being equal in each. The more teeth, the more power, provided the thickness, speed, and feed are equal. There is, however, a limit, or a point where a few teeth will not answer the place of a larger number. The thinner the saw, the more teeth will be required to carry an equal amount of feed to each revolution of the saw, but always at the expense of power. When bench-saws are used, and the sawing is done by a gauge, the lumber is often inclined to clatter and to raise up the back of the saw, when pushed hard. The reason is that the back half of the saw, having an upward motion, has a tendency to lift and raise the piece being sawn, especially when it springs and pinches on the saw, or crowds between the saw and the gauge; while the cut at the front of the saw has the opposite tendency of holding that part of the piece down. The hook or pitch of a saw-tooth should be on a line from $\frac{1}{4}$ to $\frac{1}{5}$ the diameter of the saw : a $\frac{1}{4}$ pitch is mostly used for hard, and a $\frac{1}{5}$ for softer timber. For very fine-toothed saws designed for heavy work, such as sawing shingles, etc., even from soft wood, $\frac{1}{4}$ pitch is best.

SAW-TEETH.

SAW-TEETH, Shapes of.—No. 1 is a good-shaped tooth for very soft wood, the wide bevel being the front of the tooth. The point would be liable to break or bend in very hard wood or in knots. No. 2 will stand to saw the hardest timber or knots, but will not cut as easily as No. 1. No. 3 is a form of point generally used for promiscuous sawing of both hard and soft wood. The set must be wide enough to clear the plate. After filing your saw, place it on a level board and pass a whetstone over the side of the teeth until all the wire-edge is off them. This will make the saw cut true and smooth, and it will remain sharp longer. The saw must be set true with a saw-set.

SCRAPERS, To make.—Old files which have never been recut make excellent scrapers.

SCRAPERS, Use of.—All work should be fitted as nearly true as possible before being scraped with the flat scraper, which is intended for flat surfaces only. For hollow work, curves, etc., the half-round scraper is the best, the three-cornered being the least efficient of all scrapers.

TAPS AND REAMERS, Finishing.—The forgings are got out in the usual way, left to anneal, centred, and turned just sufficient to remove the scale. Then anneal again, and turn down to within $\frac{1}{32}$ inch, or less, of finishing size. Anneal once more, and finish in the lathe. If not sprung in turning, the tap or reamer will come out all right when hardened. This has been tested successfully with taps from $\frac{1}{4}$ inch in diameter and 3 inches long up to those of 1 inch in diameter and 2 feet long.

TAPS, Tempering and hardening.—To harden a tap, heat to a cherry-red, in a clear and not a blazing fire, or, what is better, heat in charcoal, holding the tap by the square end ; dip it endwise in the water, immersing the whole of the threaded part first, and holding it still until the plain part is of a very low red ; slowly immerse the remainder, holding it still, when fully immersed, until it is quite cold. Then brighten the flutes, and temper as follows : Heat a piece of tube (with a bore about twice the diameter and a length one half that of the tap) to a bright red heat, take it from the fire, set it up vertically, and hold the tap in the centre of the tube, with the plain part of the tap in the tube and the thread part projecting. Revolve it in this position sufficient time to heat the plain end about as warm as the hand can bear it; then keep revolving the tap and moving it endwise back and forth through the tube until the thread will color to a deep brown and the shank to a brown purple. If any unevenness appears in the color, hold the parts having the lighter color a little longer in the tube, or cool the part coloring too deeply by applying a little oil to it. Perform the whole process slowly, taking the tap from the tube to retard it, if necessary.

TAPS, Tempering.—The squares of taps should be tempered to a blue.

TAPS, To temper, without springing.—Forge the tap with a little more than the usual allowance, being careful not to heat too hot, nor to hammer too cold. After the tap is forged, heat it and hold it on one end upon the anvil. If a large one, hit it with the sledge; if a small one, the hammer will do. During this operation, the tap will give way on its weakest side and become bent. Do not attempt to straighten it. On finishing and hardening the tap, it will become perfectly straight.

TOOLS, Spring of.—To obviate the spring of tools which must, of necessity, be held out a long way from the tool-post, the fulcrum off which the tool springs must be adjusted so as not to stand in advance of the cutting edge of the tool. In the engraving, a represents the fulcrum off which the tool takes its spring; B is the work to be cut; and the dotted line, C, is the line in which

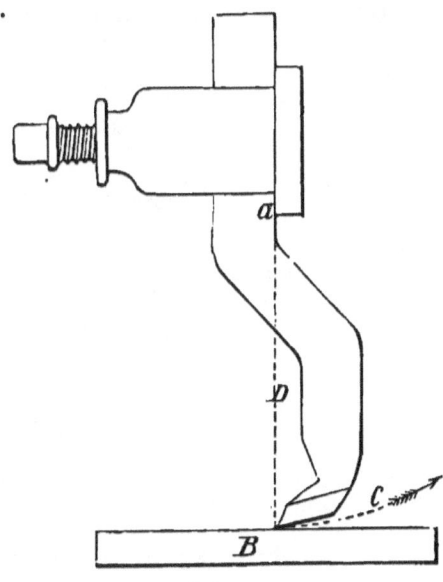

SPRING-TOOL.

the point of the tool would spring (being in the direction denoted by the arrow), which is not in this case into the cut, but rather away from it, in consequence of the point of the tool standing back from a line perpendicular to the line of the back part of the tool, as shown by the dotted line, D.

MECHANICAL SHOP-WRINKLES AND DIRECTIONS.

ANNEALING IRON AND STEEL.—Save the scales from the forge (oxide of iron) for this purpose.

ANVILS, Tempering.—This can be done by simply heating the anvil and immersing it in a tank of cold water to a depth of two or three inches. Play a stream of water from a hose on the face.

AXLES, Value of hollow.—An old apprentice of Sir William Fairbairn writes : " For several years it has cost me five dollars a week to keep the bolts on my trip or cushioned hammer-heads in repair, and, finding it to wear on my patience, I tried all kinds of iron, but to no use ; break they would. I finally bored a hole, one third the diameter of the bolts (1¼ in.), and put a ⅜ in. hole down, some way below the thread, which formed a tube. I have now run them for three months, and they show no signs of giving out. The wrench used would break the other bolts easily ; but it can not do so with these. My work on spindles requires the dies to snap together about nine times in ten, which tells very severely on the bolts."

BENDING COPPER PIPES.—Fill them with resin or lead, which will prevent them from buckling in the bends.

BOLTS AND STUDS, Standing.—When these are placed in positions liable to corrode them, they should have the standing ends ⅛ in. larger than the end receiving the nut, and the plain part should be square. By this means a wrench may be applied to extract them when necessary. The stud, also, is not so likely to break off in consequence of weakness at the junction of the thread and the plain part, where the groove to relieve the termination of the thread is cut.

BOLTS OR STUDS, Standing, To unscrew.—Studs that have become so corroded in their holes that they are liable to break off, should be well warmed by a red-hot washer ; because the strength of the stud increases by being heated up to about 400° Fahr., and therefore studs which readily twist off when cold will unscrew when heated to about that temperature. Nuts upon standing bolts of studs, in the smoke-boxes of locomotives, or in similar positions, which have become so corroded as to endanger twisting off, should be cut through on one side with a cape or cross-cut chisel, thus saving the stud at the expense of the nut. The split must be cut from the end face of the nut to the bedding face.

BRASSES, Bedding down.—In bedding down brasses or journal-boxes of any description, the following plan should be employed to gauge how much requires to be chipped or filed away from any part or parts of the bed of the brass to enable it to bed to its seat all over : Take ordinary red marking (which should be composed of Venetian-red and common oil, mixed to the consistence of a thick paint), and paint over the whole bed into which brass is to be fitted. Then take some putty (mixed stiffly), and rolling it into pellets about the size of a large pea, place them here and there upon the painted surface. Then drive the brass home, and drive it out again, when it will be found that the thickness to which the pellets have been smashed by the bottom of the brass registers to the greatest possible exactitude how near the bottom, of the brass comes to the bed of the bottom of the bearing, indicating precisely the amount to be chipped and filed off the bottom of the brass to bed it. It is better, however, to be careful not to take too much off at first, and to repeat the process with the pellets. It must be borne in mind carefully to replace the old pellets with new ones at each trial, otherwise you will be misled. The object of painting the bottom of the box with red marking, before placing the pellets, is to cause the latter to stick to the box and not to the brass, and to prevent them from falling out when the box is turned upside down to drive the brass out. This is the only practical method of ascertaining how much to take off a brass to bed it, and will be found an easily applied and almost invaluable assistance.

BRASSES, Fitting.—In doing this, a piece of wood must be used to hammer on in driving them in and out, since driving them with the bare hammer, a piece of metal, or a mandril, stretches the skin and enlarges the diameter across the bore ; then when the brasses are bored and the stretched skin is consequently removed, the brass resumes its original shape, and hence becomes loose in the strap or box. In fitting brasses to their places, leave them a little too tight, since all brasses contract a little across the bore in consequence of the process of boring. This rule applies also to journal-boxes of cast-iron or any other metal.

BRASSES, Setting.—In setting brasses or any other journal-boxes to be bored. place a piece of sheet-tin between the joint of the brasses, and bore the brasses or boxes the thickness of the tin too large, which thickness may be gauged by placing a small piece of the same tin under the leg of the inside calipers when trying the bore. The reason for this is that practice demonstrates it to be an invariable rule that a half-circle or half-hole, whether in a movable brass or in a solid box, will never fit down upon its journal, but will bind upon the edges across the diameter, and must therefore be scraped or filed on the sides to let the crown down. This defect is obviated by the employment of the sheet-tin as described, which will save three fourths of the time usually required to fit such work to a good bearing. This plan is highly advantageous in boring eccentric straps and large brasses ; and the larger the size, the thicker the tin may be.

BRASS TURNINGS AND FILINGS, To melt, with little waste.— Compress in a crucible until it is full, then cover, and lute the top with pipe or fire-clay.

BURR STONES. Filling holes in.—Use melted alum mixed with burr-stone pulverized to the size of grains of sand.

BURRS, To prevent heating,—Dress from centre to circumference, leaving no bosom. Draw a line across the centre, each way, dividing a four-foot burr into 16 squares or divisions, and other sizes, more or less, in the same proportion, with all straight furrows. Let the draft be $\frac{1}{4}$ the diameter of the rock. Lay off the lands and furrows $\frac{1}{4}$ inch each, observing to dress smooth. Sink the furrow at the eye $\frac{1}{4}$ inch deep for corn, and run out to $\frac{3}{16}$ at the periphery; for wheat, $\frac{3}{16}$ at the eye, and $\frac{1}{8}$ at the periphery. When thus finished, crack the lands in straight lines, square with the draft of cross-lines, so as to make the lands face in the runner and bed direct. This will never heat.

CARPENTER'S BENCH, To make a.—Take three pieces of 2 x 5 in. stuff, 3 feet long, for supports for top. Take two 12 in. boards, 12 feet long and 1 in. thick, for sides; nail the sideboards firmly on to the ends of the 2 x 5 cross-pieces, and put on a top of suitable material, and you have a bench without legs. Then take four pieces of 2 x 5 in. stuff of the desired height for the legs, and frame a piece 1 x 3 inches across each pair of legs, 6 inches from the bottom of the leg, putting the legs at the proper distance apart for width of bench. Cut a fork or slit in the top end of each leg, so as to straddle the cross-piece at the ends; put a $3\frac{1}{2}$ x $\frac{5}{8}$ in. bolt through each leg and the side-board, and you have a good solid bench, that can be taken down in five minutes by simply removing the four bolts. It can also be taken through any door or window, or down or up stairs, or to any place required.

CASTINGS, BRASS.—These shrink $\frac{1}{8}$ inch to the foot in cooling.

CASTINGS, COPPER, To prevent air-holes in.—Cast in green sand, and not in dried moulds, using $1\frac{1}{4}$ lbs. zinc as a flue, as pure copper will not run sufficiently freely to prevent honeycombing.

CASTINGS, COPPER.—These shrink $\frac{1}{4}$ inch to the foot in cooling in the moulds, and also shrink (as do all other castings) most where there is the greatest substance of metal.

CASTINGS, Holes in, To prevent.—In casting iron on iron or steel spindles, the moulds are cast endwise, letting the cast metal covering the spindle be an inch longer on the uppermost side than is necessary when the job is finished; thus the air-holes, if any, will form in the extra inch of length, and may be cut off in the lathe.

CASTINGS, IRON.—These shrink $\frac{1}{10}$ inch to the foot in cooling in the moulds.

CASTINGS, Shrinkage of.—Shrinkage sideways and endwise, on castings of 3 inches and less size, is compensated for by the shake in the sand given by the moulder to the pattern in order to extract it from the mould.

CASTINGS, Small.—In very small castings requiring to be of correct size, allowance should be made in the pattern for the shake of the pattern in the sand, thus: A pattern of an inch cube will require to be made $\frac{1}{32}$ inch less endwise and sideways, and

the usual allowance above an inch must be made on the top face
of the pattern, which should have " top" marked on it ; thus the
shake endwise and sideways given to the pattern, in order to extract it without lifting the sand, will be allowed for in the size
of the pattern. The effect of this shake in the sand is appreciable in patterns up to about four inches diameter. It is a common practice to cool brass castings in order to loosen or remove
the sand from the surfaces. The effect is to place conflicting
strains and tensions upon the whole skin of the metal, which will
alter its shape when the outer skin of such faces is removed ; so
that, supposing the casting to be a cube, no one face will be
either true of itself or with the others when it has been planed,
no matter how true the surfacing may have been performed.

CASTINGS, Smooth.—A means highly recommended for obtaining very smooth castings, is mixing with the green foundry sand
forming the mould about $\frac{1}{70}$ part of tar, a mixture which is employed without the addition of any other substance.

CASTINGS, To estimate the weight of.—

A pattern weighing 1 lb., and made of	Will weigh when cast in				
	Cast Iron.	Zinc.	Copper.	Yellow Brass.	Gun Metal.
Mahogany	8	8	10	9.8	10
" (St. Domingo)	10	9.5	12	11.5	12
Maple	10	9.8	12.5	12	12.4
Beech	11	11	14	15.4	13.8
Cedar	11.5	11.4	14.5	14	14.5
Yellow Pine	13	12.6	16	15.5	16
White Pine	14	14.5	18	17.5	17.8

Example : The pattern of a wheel made of cedar weighs 8 lbs,
how much will a casting of iron weigh ? 8 lbs. weight of pattern multiplied by 11.5, which is the multiplier for iron set opposite cedar, equals 92.0 weight of casting.

CHUCK-CEMENT, Removing, from lathe-work.—Warm the object over a spirit-lamp, and tap lightly with a stiff brush ; the
wax will adhere to the latter. If in a hurry, a few seconds' boiling in alcohol will remove the remainder of the wax.

CHUCK-LATHE, An electric.—In order to obviate the inconve-
nience and loss of time involved in the ordinary mode of fixing
upon a lathe-chuck certain special kinds of work, such as thin
steel disks or small circular-saws, the chuck may be converted
into a temporary magnet, so that the articles, when simply placed
on the face of the chuck, are held there by the attraction of the
magnet ; and, when finished, can be readily detached by merely
breaking the electric current and demagnetizing the chuck. The
face-plate of the magnetic chuck is composed of a central core of
soft iron, surrounded by an iron tube, the two being kept apart
by an intermediate brass ring ; and the tube and core are each

surrounded by a coil of insulated copper wire, the ends of which are connected by two brass contact-rings that encircle the case containing the entire electro-magnet thus formed. These rings are grooved, and receive the ends of a pair of metal springs connected with the terminal wires of a battery, whereby the chuck is converted into an electro-magnet capable of holding firmly on its face the article to be turned or ground. For holding articles of larger diameter, it is found more convenient to use an ordinary face-plate, simply divided into halves by a thin brass strip across the centre ; a horseshoe-magnet, consisting of a bent bar of soft iron, with a coil of copper wire around each leg, is fixed behind the face-plate, each half of which is thus converted into one of the poles of the magnet. The whole is inclosed in a cylindrical brass casing, and two brass contact-rings, fixed around this casing, are insulated by a ring of ebonite, and are connected with the two terminal wires of the magnet-coils. A similar arrangement is also adapted for holding work upon the bed of a planing or drilling machine, in which case the brass contact-rings may be dispensed with, and any desired number of pairs of the electro-magnetic face-plates are combined so as to form an extended surface large enough to carry large pieces of work. For exciting the electro-magnet, any ordinary battery that will produce a continuous current of electricity can be used.

COCKS, To grind plugs in.—The best material for this purpose is the red, well-burnt sand from the core of a brass casting, the sand being sifted, before using, through fine wire gauze. It will cut more truly and smoothly than any other material, and should be used with water, and not oil. After the cock is sufficiently ground, wash it and the barrel with clean water ; and after wiping them comparatively but not quite dry, revolve the plug backward and forward in its place under a slight pressure, and the surface will assume a rich brown color, showing very distinctly the precise nature of the fit, and leaving a smooth surface, free from grit and not liable to either jam or abrade. Valves may be ground to their seats in a similar manner.

CONES IN HOLES, Fitting.—Rub the cone with either red marking or else chalk, giving it in either case a very thin coat. A narrow strip of marking, running from end of the cone, will do. Put the cone into its place, press it forward and revolve it back and forth, and the marks will show where it binds.

CONNECTING-RODS, Fitting.—When the cross-head and crank-pin are in their places upon the engine, fitting may be done as follows : Connect the cross-head end of the rod in its place upon the cross-head journal, keeping the other end clear of the crank-pin ; put the brasses and keys in their places in the rod end, then lower the crank-pin end upon the crank-pin journal, which will show whether the cross-head journal leads true ; if it does not, move the crank-pin end of the rod back and forth, exerting a side pressure on it in the direction in which it wants to go, so as to plainly mark where the connecting-rod brasses of the cross-head end require easing ; and after the cross-head end is adjusted, put the crank-pin of the rod upon its place upon the crank-pin, keeping the cross-head end clear of the cross-head ; put the brasses, keys, etc., in their places, and proceed as before. Red marking should be rubbed on both the crank-pin and cross-head journals, so as to mark the brasses plainly. A half-round file and half-round scraper should be used to adjust and ease the brasses.

CONNECTING-ROD STRAPS, To close the jaws of.—If the jaws are too wide at the points, rest the strap upon the two ends, and (with a round pene-hammer) lightly hammer the outside of the crown of the strap all over, taking care to hammer it evenly, so as not to leave any deep hammer-marks.

CONNECTING-ROD STRAPS.—To open the jaws of a connecting-rod strap, hammer the inside face of the crown of the strap lightly and evenly all over with a round pene-hammer. To bring back to its original shape a strap that has been opened or closed in its width between the jaws, by being pened with a hammer, remove by filing $\frac{1}{32}$ inch in depth of the surface that has been hammered, or heat the part hammered to a low red heat.

CONNECTING-ROD STRAPS, Refitting, to rods.—Place the gib and key in the strap to prevent it from warping, and heat the crown end of the strap to a bright red ; on taking it from the fire, remove the scale by lightly filing with a coarse file ; then plunge the strap to nearly half the thickness of the crown in water, and after holding it there for about ten seconds, slowly immerse the remainder of the strap in the water, and withdraw when quite cold. It will be found to have closed along its whole length, although mostly at the points.

COPPER OR BRASS, To soften.—Heat to a low red heat, and plunge in salt water.

COPPER, To harden.—Copper may be slightly hardened by closing the grain. This can be done by lightly hammering its surface with a round pene-hammer.

CORUNDUM-WHEELS, To true.—The wheel being adjusted in the lathe, revolve it very fast, holding a piece of corundum-stone against the uneven or wabbling surface. In a short time, the piece will melt and unite itself to the wheel, so as to make the latter perfectly true.

COTTON-WASTE, To clean greasy.—Boil it in a strong solution of common soda in water, and save the resultant soapy liquid to keep your drills and reamers wet when boring iron.

CRANK-PINS, Riveting.—In riveting over the end of a crank-pin, apply the hammer most around the outer circumference and least toward the centre ; this will prevent the riveted end from splitting.

EMERY, for grinding purposes, To prepare fine.—When no fine emery is at hand, place coarse emery upon an iron block or plate, and hammer it well with the face end of the hammer ; after which, grind it by abrasion with the hammer face.

EMERY-WHEELS, Speed of.—A 6-inch emery-wheel should

make about 2400 revolutions per minute ; an 8-inch, 1800 ; a 12-inch, 1200.

ENGINE-ROOM CHAIRS.—Good chairs for engine-room or factory use are easily made of light gas-pipe.

GREASE, To clean, from bolts.—Moisten in benzine, roll in sawdust, and brush afterward.

GRINDSTONES, Care of.—These should never be left in the sunlight, or with one side standing in water, since heat evaporates the water in the stone, leaving it harsh and hard, while saturation softens it. The grindstone is a self-sharpening tool ; and after being turned in one direction for some time (if a hard stone), the motion should be reversed. Sand of the right grit applied occasionally to a hard stone will improve it.

GRINDSTONES, Selecting.—When you get a stone that suits your purpose, send a sample to the dealer to select by; a half-ounce sample is enough, and can be sent by mail.

GRINDSTONE-SPINDLES.—Common grindstone-spindles, with a crank at one end, are open to the objection that the stone will never keep round, because every person is inclined, more or less, to follow the motion of his foot with his hand, which causes the pressure on the stone to be unequal. The harder pressure is always applied to the very same part of the stone, and will soon make it uneven, so that it is impossible to grind a tool true. To avoid this, put in place of the crank a small cog-wheel to the spindle, say with twelve cogs ; have another short spindle, with a crank and a cog-wheel of thirteen cogs, to work into the former. The stone will make about 0.07 of a revolution more than the crank, and the harder pressure of the tool on the stone will change to another place at every turn ; and the stone will keep perfectly round if it is a good one.

GRINDSTONE, To true a carpenter's.—Use a ¾-inch bar of iron, or a gas-pipe, for a turning tool, held below the centre of the stone.

HARDENING, To prevent cracking of metals while.—Heat the water as hot as bearable to the hands, dip the metal endwise, and immerse with the thickest part of the metal downward. When fully immersed, hold the metal still until it is quite cold.

IRON, CAST, To harden.—In 3 gallons of clean water, mix ¼ pint oil of vitriol and 2 ozs. saltpetre. Heat the iron to a cherry-red, and dip as usual.

IRON, CAST, Mixture for cast-iron cylinders, requiring to be hard.—Twenty pounds charcoal pig No. 5, 40 lbs. Scotch pig, 300 lbs. scrap-iron.

IRON, CAST, Mixture for strong and close-grained cast-iron for steam cylinders, etc.—Eighty pounds charcoal pig No. 5, 100 lbs. Scotch pig, and 100 lbs. scrap-iron.

IRON, MALLEABLE, CAST, To harden.—Mix equal parts of common potash, saltpetre, and sulphate of zinc, and use as directed for prussiate of potash.

IRON PLATE, To straighten a curved.—Hammer it lightly with a round pene-hammer on the hollow side.

IRON PLATES with true, flat surfaces.—To make a true surface-plate, it is necessary to plane up three plates, which we will term A, B, and C. First fit the faces of A and B together as nearly as possible with a smooth file (using a hardly perceptible coating of Venetian-red and common oil applied to A, to show by the marks where the faces touch). Then file C to fit A. Then try C with B ; and if they mark all over the faces, they are true, and the surfaces may be finished by the scraper, trying them repeatedly as above. If, however, C and B should be found to fit on the edges only, it would demonstrate that A and B, though fitting, are not flat surfaces ; but that A is hollow and B rounding. Having corrected these defects as nearly as possible, apply B to C, again repeating the correcting process until all three surfaces, applied interchangeably, mark evenly all over, when the surface of each will be practically true. It must, however, be borne in mind that, after rubbing the surfaces together to test them, the middle of each plate will (from having received the most abrasion) show the marks the plainest, so that the plates will be more nearly true if the marks show a little the plainest at and near their edges, and less plain toward the centres. The back of each plate should be planed off, care being taken that it rests evenly upon the bench, so that neither plate shall deflect from its own weight, as it would do if its weight were not supported evenly upon the face resting upon the bench. The scraper should be used dry upon cast-iron, and kept moistened with water for steel, wrought-iron, and brass.

IRON, RED-HOT, To mark measures on.—Blacksmiths frequently measure a piece of iron, and put chalk-marks where they desire to cut it. The iron may then be placed in the fire and heated to a bright red without effacing the marks.

IRON, To remove hard skin from.—Hard skin on iron is due to heat and friction, and may be removed by heating to a dull red, filing the surface, and putting the iron to cool in lime or ashes.

IRON, WROUGHT, Contraction of.—Wrought-iron may be made to contract to a slight extent by heating it to a low red heat, and quenching it in water. The first operation only, however, is effective. This plan is used to shorten eccentric rods, etc., to avoid getting them out of true by upsetting them with blacksmiths' tools.

IRON, WROUGHT, To close holes in.—If a washer or other piece of wrought-iron is a trifle too large, fill the hole and part of the outside faces with fire-clay ; heat the iron to a bright red, and plunge it in cold water. The contraction of the circumferential iron will squeeze the metal round the hole (which has been prevented from cooling so rapidly by the clay) inward, diminishing the size of the hole. To refit a bolt to a hole in which it has worn a trifle loose, case-harden it by the prussiate of potash process, which will increase the diameter of the bolt. If it fits into a hole of wrought-iron or steel, that too may be case-hardened, which will diminish its size, and thus refit it to the bolt.

JOINTS, RUBBER.—In making a rubber joint, take a piece of chalk and rub it on the side of the rubber and flange where the joint is to open ; and when required, they will come apart easily, and not break the rubber, although the latter may be burnt and hard.   Repeat the chalking before screwing up, and you will have as good a joint as ever, and the rubber can be used a great number of times.

KEYS, Driving.—Drive the key into the keyway to mark where it binds.   The keyway should be oiled first, especially if the metal is cast-iron ; otherwise the surfaces are liable to seize a cut, making it very difficult to drive the keys out, and cutting ragged grooves in both the keyway and the key.   The same rule applies to crank-pins and all similar work.

KEYS, To make.—These should be made to fit the keyways at a good fit on the sides without being tight, the locking being per-formed by the taper of the top and bottom, the amount of which should be about $\frac{1}{8}$ inch per foot of length.

KEYWAY AND SLOTS, To ease, when hardened.—Take a strip of copper and use it as a file, applying oil and fine emery upon the surface of the work.

LATHE, Setting work on the face-plate of a.—Let the work be set out and first lightly prick-punched ; then clamp to place light-ly as near as possible, but never set the "dead-centre" against the work, for that will not bring it true ; now with a scratch-awl or sharp-pointed centre, with the point resting in the prick-mark, and the other end held against or on the dead centre, revolve the work.   If the point marked for the centre of the hole is out of true, the scratch-awl, or whatever rests in the point, will vibrate. Put into the lathe-rest a tool, without fastening it, and push it up to the scratch as the work is revolved, and the extent of the vibration can be seen.   The work can be driven as thus indicat-ed when there is no vibration of the scratch or centre, the work is perfectly set, and may be securely fastened.

LINERS, Thickness of.—To ascertain the proper thickness of a liner or strip necessary to go between a pair of brasses so that (when the faces do not meet) the liner may be placed between them and the brasses, when bolted up tight, without jamming the journal, place a piece of lead wire between the brasses, and then screw the cap down tight, and the lead wire will compress, denoting the necessary thickness of liner.   The latter should be made a shade thicker than the distance the wire was compress-ed, so that the brasses may fit without binding the journal.

LINKS, To close a quadrant or link that has sprung or opened in hardening.—Clamp it with bolts and plates, placing the die in the slot to support any part which does not require to be closed.   *To open the slot of a quadrant or link that has closed in being hardened :* Take two keys having an equal amount of taper upon them, and place them together so that their outside edges are parallel.   In-sert them in that part of the slot which requires to be opened, and holding a hammer against the head of one key on one side of the link, drive in the other key with a hammer on the other side of the link.   After the key is driven as far in as the judgment suggests, measure the width of the slot, so that, if the operation was not performed to a sufficient extent on the first attempt, the operator may judge how much to essay at the second, and so on. *To prevent, as far as possible, a slot link or quadrant from altering its shape in the process of hardening :* Fit into the slot, at various parts along its length, pieces of iron of the same diameter as the die intended to work in the slot, and in quenching the quadrant, immerse it endwise and vertically.

NAILS into hard wood, To drive.—Dip in grease to assist penetration.

NUTS, Tight, To ease.—To ease a nut that is a little too tight upon a bolt, place it upon the bolt, and, resting it upon an iron anvil or block, strike the upper side with a hammer, turning the nut so that not more than two blows will fall upon the same face. The smaller the nut, the lighter the blows should be, and *vice versa.*

NUTS, Tight, To unscrew.—To start a nut that is corroded in its place, so that an ordinary wrench fails to move it, strike a few sharp blows upon its end face ; then holding a dull chisel across the chamfer of the nut, strike the chisel-head several sharp blows, which will, in a majority of cases, effect the object.

PATTERNS, Cast-iron.—These should have their surfaces smoothed, and be then slightly heated and waxed all over with the best beeswax.

PATTERNS FOR BRASS BED-PIECES.—In making a pattern for a brass bedding in a circular bed, first make the pattern at the same curve as the bed, and then pare off the centre of the crownbed about $\frac{1}{64}$ inch for every inch of diameter of bore of the brass ; the reasons for this are explained in treating of patterns for semi-octagonal bedding-brasses.

PATTERNS FOR BRASSES.—In making a pattern for a brass to fit in a semi-octagonal bed such as is employed in pillar-blocks, and sometimes in the small ends of connecting-rods and axle-boxes : after having made the bed of the brass to the same shape as the seat into which it beds, take off $\frac{1}{16}$ inch in brasses below 3 inches bore, or $\frac{1}{8}$ inch in brasses above that size, from the crown face of the brass pattern, for the following reasons : The casting of iron or of brass contracts, in cooling, most at the sides, and the above is to compensate for this contraction. Furthermore, it will require only $\frac{1}{16}$ inch to be cut off the angles to let a brass (having bed-angles at 40°) down $\frac{1}{8}$ inch on the crown ; whereas it will require $\frac{1}{8}$ inch taken off the crown face to let the bed-angles down $\frac{1}{16}$ inch. A strict observance of this rule will, in all cases, save one half the time required to fit such brasses to their places. In brasses whose bed-angles are more acute, a greater allowance must be made.

PATTERNS, To fit.—To get a pattern to fit closely over an irregular casting having angles, projections, or crooks in it (such as is often required to make a casting with which to patch cylinders or junctions of pipes), take a piece of sheet-lead, and hammer it lightly with a round pene-hammer, closing it round the casting until it will, by stretching where it is requisite, conform

strictly to the shape of the surface, however irregular it may be.
The moulder can then cast a patch from the sheet-lead, making
it of any required thickness.

PATTERNS, Wooden.—These should never be left in the foundry, where they are liable to warp from the excessive range of
temperature.

PENING, Setting work by.—The operation termed "pening," is
stretching the skin on one side of work to alter its shape, the
principle of which is that, by striking the face of the metal
with a hammer, the face of the metal struck stretches, and
tends to force the work in a circular form, of which the
part receiving the effect of the hammer is the outside
circle or diameter. The engraving represents a piece of

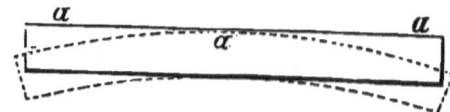

flat iron, which would, if it were well hammered on the face,
a a a, with the pene of a hammer, alter its form to that denoted by the dotted lines. Or let the rod, *a,* shown below be
attached to a double eye at one end, the other end requiring to
come fair with the double eye, *b,* at the other end; then, if it is
pened perpendicularly on the face, *c,* of the rod, the stretched
skin will throw the end around so that it will come fair with the

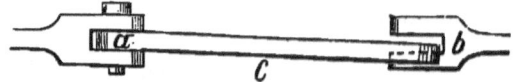

eye, *b.* Connecting-rod straps which are a little too wide for the
rod ends may be in like manner closed so as to fit by pening the
outside of the crown end, or, if too narrow, may be opened by
pening the inside of the crown end; but in either case, the ends
of the strap alter most in consequence of their lengths, and the
strap will require refitting between its jaws.

PIPES, GAS OR STEAM, Threads used in screwing:

Inside diameter.	Threads per inch.	Inside diameter.	Threads per inch.
$\frac{1}{8}$	27	$1\frac{1}{4}$	$11\frac{1}{2}$
$\frac{1}{4}$	18	2	$11\frac{1}{2}$
$\frac{3}{8}$	18	$2\frac{1}{2}$	8
$\frac{1}{2}$	14	3	8
$\frac{3}{4}$	14	$3\frac{1}{2}$	8
1	$11\frac{1}{2}$	4	8
$1\frac{1}{4}$	$11\frac{1}{2}$		

Taper of threads, $\frac{1}{16}$ per inch of length. These inside diameters are
only estimated, as they vary for pipes of different strength, the

thickness varying for the different grades, and the outside diameter remaining the same.

PISTON-RINGS, To open.—Hammer them lightly with a round pene-hammer all round their inside faces.

PISTON-ROD GLANDS.—If these are rather tight, the piston-rod may be eased by rubbing the gland up and down the rod, and giving it at the same time a revolving motion back and forth at each stroke. Oil must be supplied to the rod to prevent the gland from seizing or cutting. A gland should be chucked in the lathe by the flange, so that the bore and outside diameter may be turned at one chucking, and thus be true without depending upon the truth of a mandrel.

PLASTER, Inserting screws in.—Make a large hole in the plaster and drive in a wooden plug, in which insert the screw. The plug may be first split and a groove cut in each half.

PULLEYS, Turning.—Pulleys should be turned either on a mandrel, or else chucked by the arms, since chucking them by the rims springs them out of true.

PUNCHING METALS.—The same elements of resistance enter into the operation of punching as in that of shearing. In short, a punch and die may be considered as shears with circular blades. The coefficient of pressure in punching, for any given area of section, will be exactly that for shearing the same area of sec-

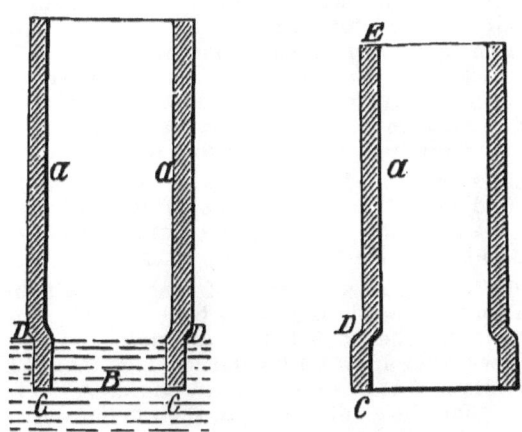

SHRINKING METAL-WORK.

tion, without reference to the thickness of the material. The measure of force necessary to effect the various punchings readily gives the value of the resistance to shearing in cases of ordinary metals. This resistance, per square foot, is determined to be, in lbs.: For lead, 392,548 ; block-tin, 450,784 ; alloy of lead and tin, 731,176 ; zinc, 1,843,136 ; copper, 4,082,941 ; iron, 103,-333.

REDUCING THE SIZE OF WORK BY SHRINKING WITH FIRE AND WATER.—For reducing the size of wrought-iron work, the process shown in the following engraving may be employed. *a a* is the section of a wrought-iron square box or tube, which is supposed to be made red hot and placed suddenly in the water, B, from its end, C, to the point D; the result is that the metal in the water, from C to D, contracts or shrinks in diameter, and compresses the hot metal immediately above the water-line, as the small cone at D denotes. If then the box or tube is slowly immersed in the water, its form, when cold, will be as in the right-hand figure, that part from C to D maintaining its original size, and the remainder being smaller.

It must then be reheated and suddenly immersed from the end, E, nearly to D, until it is cold, and then slowly lowered in the water, as before, which will contract the part from D to C, making the entire length parallel but smaller, both in diameter and bore, than before it was thus operated upon.

ROLLS, To prevent cinders getting between the necks of.—Bore grooves out of the bearings, $1\frac{1}{4}$ inches wide and $\frac{1}{4}$ inch deep and $1\frac{1}{2}$ inches apart, put them at an angle of 45° with the face of the brass, and fill up said grooves with soft Babbitt metal. Then when cinder or iron gets in, it will travel but a short distance before it reaches the soft metal, and the motion of the roll will imbed it therein so that it can not protrude and score the neck, as it would were it to stick in the brass.

RUST, To remove, from small hollow castings.—Dip in dilute sulphuric acid 1 part commercial acid to 10 water; wash in hot lime-water, and dry in the tumbler with dry sawdust.

SANDS AND FACINGS FOR CASTINGS.—For castings, such as pipes or small cylinders, fine sand, termed No. 1, is used, the facing being plumbago. A good facing for loam castings is made of 1 part Whitehead sand to 3 parts fire-sand. For very fine-faced castings, Albany or Waterford sand is unsurpassed. Another facing for fine castings is 1 part of sea-coal to 8 or 10 of Albany sand ; for heavy castings, however, 1 to 5 will answer.

SAW-BLADES, Small.—Mechanics who want small gig saw blades will find that the steel springs of which hoopskirts are formed will make capital ones of any lengths; and they vary in width, so as to be suitable for a variety of uses.

SCREW-DRIVERS, The advantage of long.—The reason that a screw is driven more easily into wood by a long than by a short screw-driver, is that the tool is held at an angle, and consequently the long screw-driver affords a greater leverage than a short one. If both were secured so as to be at right angles to the face of the screw, there would be no difference in their action.

SCREWS, To prevent, getting tight in their nuts.—Plane a keyway or groove in the screw, $\frac{1}{4}$ in. wide, the full length of the screw and down to the bottom of the threads ; and it will act like a tap, and scrape all the hard gummy grease out of the nut, and always keep it clean and working free.

SCREW AND BRAD HOLES IN FINISHED WORK, To plug.—Glue the edge of the plug ; put no glue in the hole. By this means the surplus glue is left on the surface, and if the plug does not hit the screw, it will seldom show. Set the heads of brads well in, then pass a sponge of hot water over them, filling the holes with hot water. This brings the wood more to its natural position, and closes by degrees over the head of the plug. When dry, sandpaper off and paint, and the putty will not hit the head of the brad.

SCREWS, Hints about.—When screws are driven into soft wood and subjected to considerable strain, they are likely to work loose ; in such case, dip the screw in thick glue before inserting. When buying screws, see that the heads are round and well cut ; that there are no flaws in the body or thread part, and that they have gimlet points. A screw of good make will drive into oak as easy as others into pine, and will endure having twice the force brought against it. When there is an article of furniture to be hastily repaired, and no glue is handy, insert a stick a little less in size than the hole for the screw and fill the rest of the cavity with powdered resin ; heat the screw sufficiently to meet the resin as it is driven in.

SCREW, To remove.—An obdurate screw may sometimes be drawn by applying a piece of red-hot iron to the head for a minute or two, and immediately using the screw-driver.

SCREW-THREADS.—English and American proportions :

The Whitworth Thread.

Diameter in inches	$\frac{3}{16}$	$\frac{1}{4}$	$\frac{5}{16}$	$\frac{3}{8}$	$\frac{7}{16}$	$\frac{1}{2}$	$\frac{5}{8}$	$\frac{3}{4}$	$\frac{7}{8}$	1
Threads per inch	21	20	18	16	14	12	11	10	9	8

Diameter in inches	$1\frac{1}{8}$	$1\frac{1}{4}$	$1\frac{3}{8}$	$1\frac{1}{2}$	$1\frac{5}{8}$	$1\frac{3}{4}$	$1\frac{7}{8}$	2	$2\frac{1}{8}$	$2\frac{1}{4}$
Threads per inch	7	7	6	6	5	5	$4\frac{1}{2}$	$4\frac{1}{2}$	4	4

Diameter in inches	$2\frac{1}{2}$	3	$3\frac{1}{4}$	$3\frac{1}{2}$	$3\frac{3}{4}$	4	$4\frac{1}{4}$	$4\frac{1}{2}$	$4\frac{3}{4}$	5
Threads per inch	$3\frac{1}{2}$	$3\frac{1}{2}$	$3\frac{1}{4}$	$3\frac{1}{4}$	3	3	$2\frac{7}{8}$	$2\frac{7}{8}$	$2\frac{5}{8}$	$2\frac{1}{2}$

Diameter in inches	$5\frac{1}{4}$	$5\frac{1}{2}$	$5\frac{3}{4}$	6
Threads per inch	$2\frac{5}{8}$	$2\frac{5}{8}$	$2\frac{1}{2}$	$2\frac{1}{2}$

Angle of threads = 55°. Depth of threads = pitch of screws. (One sixth of the depth is rounded off at top and bottom.) Number of threads to the inch in square threads = $\frac{1}{2}$ number of those in angular threads.

Standard American Proportions.

Diameter in inches	$\frac{1}{4}$	$\frac{5}{16}$	$\frac{3}{8}$	$\frac{7}{16}$	$\frac{1}{2}$	$\frac{9}{16}$	$\frac{5}{8}$	$\frac{3}{4}$	$\frac{7}{8}$	1
Threads per inch	20	18	16	14	13	12	11	10	9	8

Diameter in inches	$1\frac{1}{8}$	$1\frac{1}{4}$	$1\frac{3}{8}$	$1\frac{1}{2}$	$1\frac{5}{8}$	$1\frac{3}{4}$	$1\frac{7}{8}$	2	$2\frac{1}{8}$	$2\frac{1}{4}$
Threads per inch	7	7	6	6	$5\frac{1}{2}$	5	5	$4\frac{1}{2}$	$4\frac{1}{2}$	4

Diameter in inches	$2\frac{1}{2}$	3	$3\frac{1}{4}$	$3\frac{1}{2}$	$3\frac{3}{4}$	4	$4\frac{1}{4}$	$4\frac{1}{2}$	$4\frac{3}{4}$	5
Threads per inch	4	$3\frac{1}{2}$	$3\frac{1}{4}$	$3\frac{1}{4}$	3	3	$2\frac{7}{8}$	$2\frac{5}{8}$	$2\frac{5}{8}$	$2\frac{1}{2}$

Diameter in inches	$5\frac{1}{4}$	$5\frac{1}{2}$	$5\frac{3}{4}$	6
Threads per inch	$2\frac{1}{2}$	$2\frac{3}{8}$	$2\frac{3}{8}$	$2\frac{1}{4}$

Angle of threads = 60°. Flat surface at top and bottom = $\frac{1}{8}$ of the pitch. For rough bolts, the distance between parallel sides of bolt-head and nut = $1\frac{1}{2}$ diameters of bolt + $\frac{1}{8}$ of an inch. Thickness of head = $\frac{1}{4}$ distance between parallel sides. Thickness of nut = diameter of bolt. In finished bolts, thickness of head = thickness of nut. Distance between parallel sides of a bolt-head and nut and thickness of nut is $\frac{1}{16}$ of an inch less for finished work than for rough.

SOFTENING BRIGHT WORK WITHOUT DAMAGING THE FINISH.— Place the pieces in an iron box, and fill in the interstices with iron turnings; close the box, lute the cracks with fire-clay, and heat to a red, allowing the box to cool as slowly as possible. It is a good plan to let the furnace-fires go out and leave the box in the furnace to cool.

SOLDER, To flow.—Ordinary solder, 2 parts tin and 1 part lead, will flow smoothly on tin when dipped by previously putting sal-ammoniac on the surface to be tinned.

SPIRIT-LEVEL, Accuracy of the.—The best length of bubble depends on the length or curvature of the tube, a short bubble being required for a tube with a small radius of curvature, and increasing regularly in proportion with the increase of the radius of curvature.

SPRINGS, STEEL.—*To find elasticity of a given steel-plate spring:* Breadth of plate in inches multiplied by cube of the thickness in $\frac{1}{16}$ in., and by number of plates. Divide cube of span in inches by product so found, and multiply by 1.66. Result, equal elasticity in $\frac{1}{16}$ in. per ton of load. *To find span due to a given elasticity and number and size of plate :* Multiply elasticity in sixteenths per ton by breadth of plate in inches, and divide by cube of the thickness in inches, and by number of plates; divide by 1.66, and find cube-root of quotient. Result, equal span in inches. *To find number of plates due to a given elasticity, span, and size of plates :* Multiply the cube of the span in inches by 1.66. Multiply the elasticity in sixteenths by the breadth of the plate in inches, and by the cube of the thickness in sixteenths; divide the former product by the latter. The quotient is the number of plates. *To find the working strength of a given steel-plate spring :* Multiply the breadth of plate in inches by the square of the thickness in sixteenths, and by the number of plates. Multiply also the working span in inches by 11.3; divide the former product by the latter. Result, equal working strength in tons burden. *To find span due to a given strength and number and size of plate :* Multiply the breadth of plate in inches by the square of the thickness in sixteenths, and by the number of plates; multiply also the strength in tons by 11.3; divide the former product by the latter. Result, equal working span in inches. *To find the number of plates due to a given strength, span, and size of plate :* Multiply the strength in tons by span in inches, and divide by 11.3; multiply also the breadth of plate in inches by the square of the thickness in sixteenths; divide the former product by the latter. Result, equal number of plates. The span is that due to the form of the spring loaded. Extra thick plates must be replaced by an equivalent number of plates of the ruling thickness before applying the rule. To find this, multiply the number of extra plates by the square of their thickness, and divide by the square of the ruling thickness ; conversely, the number of plates of the ruling thickness to be removed for a given number of extra plates may be found in the same way

SPRINGS, To reduce elasticity of.—A well-tempered bar-spring will lose much of its elastic strength by filing off a very thin scale from the surface.

STEEL, Advantage of holes punched in.—The advantage in tensile strength, when holes are drilled in steel rather than punched, is 25.5 per cent.

STEEL, CAST, To weld.—Apply powdered borax to .the weld while heating it in the fire. If the steel is made too hot, it will crack during the hammering process.

STEEL, CAST, Welding compound for.—Mix ¼ lb. saltpetre and ¼ lb. oil of vitriol in 2 gallons hard water ; heat the steel to a blood-red, and cool in the mixture before welding. Then reheat, in sand, and weld by hammering as usual.

STEEL, Fire for tempering.—In hardening and tempering steel, a clean charcoal, anthracite, or coked bituminous coal fire is required ; such as is fit for taking a forging heat on iron is entirely unfit for hardening purposes. The sulphur contained in the coal combines with the steel to form sulphuret of iron,. and ruins its texture.

STEEL, Tempering.—The colors shown at different temperatures Fahr. are as follows: Very pale yellowish, 430° ; pale straw, 450° ; yellow, 470° ; brown, 490° ; mottled brown, 510° ; purple, 530° ; bright blue, 550° ; blue, 560° ; dark blue, 600°.

STEEL, To demagnetize.—(1) Heat it to a red heat, and allow it to cool slowly. (2) Place the steel on a magnet, with the same poles touching the same poles of the magnet, and repeat the operation till total demagnetization has taken place.

STEEL, To remove blue color from.—(1) Use 1 part oil of vitriol to 10 parts water. (2) Dip the articles in a strong solution of cyanide of potassium nearly boiling. (3) Dip the article in hydrochloric acid, and quickly rinse in clean water.

TEMPERING steel for drilling rock.—Be careful not to overheat it in hardening and forging, and quench in salt water, drawing to a brown color.

TEMPERING, The color-tests for.—Says Mr. J. Richards : " Procure eight pieces of cast-steel, about 2 in. long by 1 in. wide, and ⅜ of an inch thick ; heat them to a high red heat, and drop them into a salt-bath. Leave one without tempering, to show the white shade of extreme hardness, and grind off and polish one side of each of the remaining seven pieces. Then give them to an experienced tool-maker to be drawn to seven various shades of temper, ranging from the white piece to the dark blue color of soft steel. On the backs of these pieces paste labels, describing the technical name of the shades and the general uses to which tools of corresponding hardness are adapted. This will form an interesting collection of specimens, and accustom the eye to the various tints, which will, after some experience, be instantly recognized when seen separately."

TINNING small castings.—Clean, and boil them with scraps of block-tin in a solution of cream of tartar.

WATER-WHEEL, Steps for.—No step or foot-bearing of metal is equal to one of good oak or rock-maple.

ZINC, Stamping.—In stamping sheet-zinc in dies, much waste occurs from the small difference between the melting-point and the temperature at which sheet-zinc should be stamped to get the best effect. To obviate this waste, heat the zinc by dipping in oil at the proper temperature.

ENGINEERING.

TESTING THE STRENGTH OF MATERIALS

BY PROF. R. H. THURSTON.

THE engraving which accompanies this article illustrates a very convenient, yet quite accurate, method of determining the strength of materials, which has been devised by the writer. The test-piece is made by cutting, from the piece of metal of which the strength is to be determined, a piece about 3 in.

TESTING METALS.

long and 1 in. square. At the middle of its length, a part is turned cylindrical in form and 1 in. long, with a diameter of $\frac{1}{2}$ in. if of iron, or $\frac{3}{8}$ in. if the metal is steel. The test-piece thus made is fastened in the vise, as shown in the engraving, and a long-handled wrench is attached to the projecting head. A spring-balance is secured to the end of this wrench, and the experimenter twists off the head by pulling on this spring-balance, as seen in the illustration. The balance should be capable of indicating weights of fifty pounds or more. By simply painting the scale of the balance with white-lead, or smearing it with tallow, and by springing the pointer so that it will touch the surface, a recording apparatus may be improvised which will indicate the maximum strain reached during the test.

In testing, the experimenter pulls steadily on the balance, gradually increasing the force exerted, and watching carefully, and noting the action of, the test-piece and the balance, until fracture occurs. A resistance, which is apparently quite unyielding, is felt at first ; this is suddenly observed to be succeeded by a gradually increasing distortion of the test-piece, accompanied by an increasing resistance, up to the point of the commencement of rupture. From the latter point, the resistance becomes less and less, finally ceasing when the test-piece falls apart. By conducting the operation very carefully, and noting resistances very accurately, all of the following important points may be determined :

The *limit of elasticity* is the point at which the yielding first commences. Note the reading of the balance at this point and the angle of distortion. The last quantity is the measure of the stiffness of the metal. The most rigid pieces are, of course, those which yield the least with a given amount of force. After the piece has been twisted so far as to have taken a set, the pull may be relaxed, and the distance which the piece springs back is to be noted. The *elasticity* of the metal is measured by this recoil. The *ductility* of the metal is measured by the extent of yielding which occurs before fracture takes place. The *resilience* of the metal—which is the name given its power of resisting shock—is very closely proportioned to its strength multiplied by its ductility. Therefore, to ascertain what blow would be resisted by it without its taking a set, it is simply necessary to multiply the resistance at the limit of elasticity by the amount of distortion observed within the elastic limit. The *homogeneity* of the material is indicated by the smoothness and regularity with which the metal changes in its power of resistance as the deformation progresses.

In making such a series of experiments, it is usually found best to first select a well-known and good brand of the kind of metal which it is proposed to test, and, by a set of experiments on test-pieces cut from it, to determine what, with the particular arrangement of apparatus chosen, is the resistance registered by the balance, and what are the characteristics of the metal as shown by the method here described. By a careful comparison of the behavior of the metal of which the quality is desired to be learned with this standard set of samples, the operator soon learns to judge quickly and accurately of the value of his material for any specified purpose.

As the tensile strength of a metal is usually very closely proportional to the resistance to torsion, this also enables a very satisfactory determination of the value of the metal for resisting tension to be obtained. In the autographic recording machine, built by the Mechanical Laboratory of the Stevens Institute of

Technology, these results are permanently inscribed upon a sheet of profile-paper, the pencil of the apparatus writing a diagram or curve which is a record of all the circumstances modifying the resisting power of the metal while under test. The rule being applied, the torsional, and approximately the tensile, resistance is read off at a glance, and the position of the elastic limit, the homogeneousness, the elasticity, the stiffness, the ductility, the resilience, are all found fully indicated by the diagram, and can be, at any subsequent period, shown by means of this automatically produced record. On these records, the tensile resistance is found to be about 25,000 pounds per square inch for each inch in height of the diagram.

The peculiar method of fracture here adopted is well adapted to exhibit in the surfaces of the break any peculiarity of the metal. If homogeneous, it will show a uniform and characteristic fracture ; if seamy, it will be found to have cracks extending spirally around it ; if of cast-iron, the character of the ruptured surfaces will at once reveal to the experienced eye whether the metal is fine or coarse grained, a dark foundry or a light forge iron, and whether of close or open texture. If of steel, it will be readily seen whether it is " high " or " low," whether tool steel or of the machinery grade. Whatever the character of the material, the eye, experienced in such kinds of observation, will at once detect it, while the record of the experiment, or the " strain-diagram," will give the exact data of resistances, and will be a check upon the judgment thus formed.

THE ENGINE AND ITS APPENDAGES.

CONDENSERS, Gain from the application of, to steam-engines.—
In the early days of the steam-engine, very low pressure was ordinarily employed for engines with condensers, while, on the contrary, what was considered a very high pressure was adopted for engines that exhausted into the atmosphere. Hence arose the terms high and low pressure engines, the former being engines with, and the latter without, condensers. At present, a high pressure of steam is ordinarily carried in both kinds of engines, so that the terms do not describe the two varieties as well as formerly. Many engineers prefer to class engines as condensing and non-condensing, rather than as high and low pressure ; and this classification is generally considered the more correct of the two. One who regards economy puts in a condensing engine, if he has plenty of water in the locality ; and many old non-condensing engines are being fitted with condensers, under the more enlightened engineering practice of the present time.

It may be fairly assumed that a non-condensing engine has, on an average, at least 2 lbs. per square inch back pressure on the piston. By the application of a condenser, it might be expected that there would be a negative pressure of 10 lbs. per sq. inch on the back of the piston, so that the piston pressure would be increased by 12 lbs. In this assumption, an allowance is made for the power required to work the air-pump, and the engine is supposed to be at least 75 horse-power. For an engine smaller than this, it would be better to allow an increase in the positive pressure of not more than 10 lbs. per square inch. As the condenser, by decreasing the back pressure on the piston, adds just as much to the positive pressure, it is plain that a lower pressure of steam can be used, or the steam may be cut off at an earlier point of the stroke. The gain in either case can be approximately calculated. If the gain in positive pressure produced by the reduction in back pressure be multiplied by 100, and divided by the mean effective pressure on the piston, it will give the *percentage* of gain in pressure due to the condenser.

Thus, if the mean effective pressure on the piston is 30 lbs. per square inch, the gain in pressure will be 100 times 12, or 1200, divided by 30, which is 40 per cent. Now suppose that before the condenser was attached, the steam was cut off in the cylinder at half stroke ; under the new conditions the required mean effective pressure can be obtained with a lower boiler pressure than before. Before the condenser was in use, it would be necessary to maintain a pressure in the boiler of about 58 lbs. per square inch by gauge, to give a mean effective pressure of 30 lbs. on the piston ; while with an increase of 12 lbs. in the effective pressure, by the application of the condenser, a boiler pressure of about 39 lbs. would suffice. As the weight of steam per cubic foot at 58 lbs. pressure is 0.17481 lbs., and only 0.132 lbs. at 39 lbs. pressure, there would be a saving of about 24.5 per cent in the amount of steam required to run the engine. Instead of reducing the steam pressure after attaching a condenser to an engine, it might be better to maintain the same pressure in the boiler, and cut off the steam at an earlier part of the stroke. In the case under consideration, the increase in 12 lbs. of the effective pressure would permit of closing the steam port a little before the completion of one third of the stroke ; and supposing that the clearance space in the cylinder amounts to 5 per cent of the capacity of the cylinder, the quantities of steam required per stroke, before and after the use of the condenser, would be in the ratio of 550 to 363, so that there would be a saving of 34 per cent.

The example given represents a case in ordinary practice. By varying the data, of course a greater or less amount of saving would result ; but with an engine in good condition, it is generally safe to estimate that a saving from 20 to 25 per cent of the amount of steam used, and, consequently, of the consumption of coal, will be realized by the application of a condenser. Indeed, it is not unusual for manufacturers to guarantee this amount of saving, in converting a non-condensing into a condensing engine.

B.

COTTON MACHINERY, Power required to drive.—The following are fair approximate rules : Cotton openers, 1 horse-power per 1000 lbs. cotton delivered. Cotton pickers, 3 horse-power per 1000 lbs. cotton delivered. Cotton cards, $\frac{1}{20}$ horse-power per lb.

cotton delivered per day, and, at 125 revolutions per minute, 0.12¼ horse-power. Cotton cards, best practice, $\frac{1}{40}$ horse-power per revolution per minute. Railway heads, breakers, 1 horse-power per each 10 yards per minute. Railway heads, finishers, 0.001 horse-power per revolution per minute. Drawing-frames, 0.002 horse-power per revolution per minute. Spindles, 0.005 horse-power per spindle per 1000 revolutions. Damp weather adds 10 or 12 per cent; methods of banding may make equally great variations. Looms require from 0.1 to 0.25 horse-power each. Pickers take 4 to 6 horse-power. Cloth shears from 3 to 4 horse-power.

CYLINDERS, Balancing heavy.—The cylinder, being keyed upon its axle as it is intended to run, is lifted by a tackle or crane, and lowered so that each of its journals rests upon a stout steel straight-edge placed so that its upper surface is exactly level and parallel with its fellow. These straight-edges should not only be so rigid as to suffer no sensible deflection from the weight of the cylinder, but they should be very hard and smooth, and great care should be taken to keep them free from indentations. The journals of the cylinder must also be round and polished. The cylinder can now be loaded on its lighter side, or *vice versa*, until it will remain perfectly motionless when stopped in any part of its revolution.

CYLINDERS, LOCOMOTIVE, Placing in line.—To test the accuracy of the work after the bed-piece has been permanently fixed to the boiler, clamp a cylinder to its seat on the bed-piece, and fit a wooden cross (with a pin-hole through its centre) to the bore of the cylinder at its front end ; then pass a fine strong line through the hole, and extend it back so that it shall occupy a point exactly at the intersection of the central line of the driver-axle with the vertical plane of motion of the centre of the crank-pin and connecting-rod ; draw the line taut and fasten it in this position ; then apply callipers or a gauge at the rear end of the cylinder, between the surface of the bore and the line, above and below and right and left of the line ; and if the cylinder is in line, the four distances will of course be exactly the same. It is essential that the two horizontal distances should coincide exactly, and that the central lines of the two cylinders of a locomotive should be exactly parallel with each other, but for obvious reasons the exact coincidence of the two vertical distances is not essential to the efficiency or correct working of the engine.

Instead of a wooden cross, as above mentioned, a more convenient instrument, made of metal, may be provided, consisting of four bevel gears, A, which serve also as nuts, which work four sockets, B, with threads cut on their inner ends, all neatly fitted to a light casting, E, having a fine central hole for the line, as shown. A central gear, C, works the four gears, of course all at the same time. Several sets of steel rods, D, may be provided if necessary, of different lengths, and thus render the instrument universal in its application, each set of rods serving for cylinders varying two inches, more or less, in the diameters of their bores.

To determine whether a cylinder of an old engine is in line: Remove the front head of the cylinder, the piston, the stuffing-box gland, and the cross-head ; apply the cross and line, as above directed, extending the line through the piston-rod hole in the rear head to a point exactly central with the crank-pin when the crank is at its dead point ; draw the line taut, and, if the cylinder

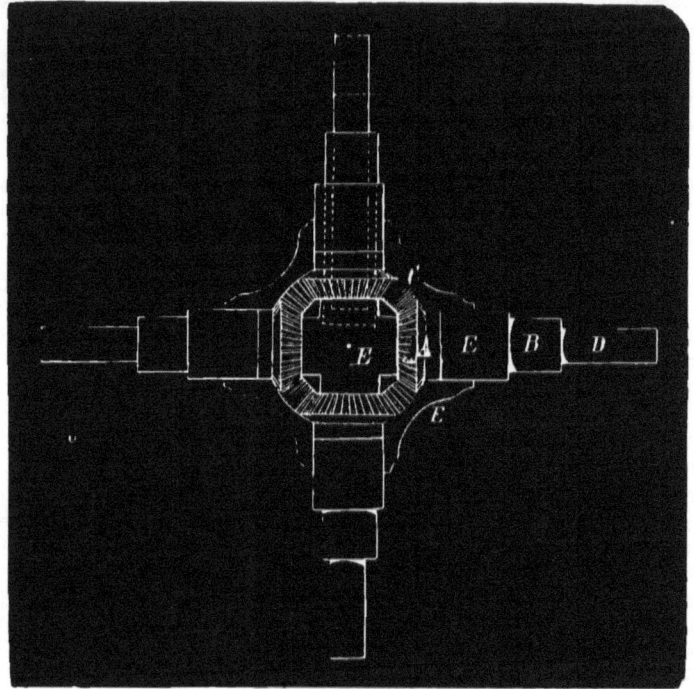

PLACING LOCOMOTIVE CYLINDERS IN LINE.

is correctly in range, the line will occupy a central position in the stuffing-box, which may be determined as before directed. If the cross-head guides are parallel with the line, both vertically and laterally, they are also correct.

CYLINDERS, Thick.—Thick cylinders are those in which the thickness is considerable in comparison with the internal diameter. To find the bursting pressure of a thick cylinder, take the product of (1) the tenacity of the material in pounds per square inch, and (2) the thickness of the cylinder in inches, and divide the product by the sum of (1) the thickness in inches, and (2) the internal radius of the cylinder in inches. Thus a cylinder with an internal radius of 4 inches, and a thickness of 5 inches, if made of cast-iron having a tensile strength of 16,000 lbs. per square inch, has a bursting pressure of 8888.9 lbs., this being the product of 16,000 and 5, divided by the sum of 4 and 5.   B.

ENGINEER, Duties of the.—The ordinary daily duties of an engineer are as follows : On coming in the morning, he should first ascertain the amount of water in the boiler ; and, if that is all right, proceed to raise steam, either cleaning and spreading the

fire, if it has been banked, or making it up, if it has been hauled.
A fire is kindled in the boiler in essentially the same manner as
in a stove, wood and shavings first being ignited, and then covered with coal. In starting the fire, it is a good plan to cover the
back of the grate with coal, to prevent the passage of cold air
through the tubes. In getting up steam, the safety-valve should
be raised a little, to permit the escape of air from the boiler.
Having got the fire under way, the engineer should wipe off the
engine, fill the oil-cups, and make any adjustments that may be
necessary, such as tightening keys and screwing up joints or
glands of stuffing-boxes, and should see that the cylinder-cocks
are open. When steam is raised, he should open the stop-valve
and start the engine ; after which, if a part of his duty is to attend to the shafting, he should examine and oil it. Then he
should get out the ashes, provide a supply of coal, and screen it,
if necessary, and proceed to make every thing tidy around the engine and boiler. Throughout the day, he should keep a watchful
eye on the fire, the water, the steam, and the engine. In manag-
ing the fire, care should be taken to have the furnace-door open
as little as possible ; and, if steam is formed too rapidly, the fire
should be regulated by closing the damper and ash-pit doors. In
regulating the height of the water, it is a good plan to keep a
steady feed, and maintain the height constant. If it is found
that the water is falling, the engineer should discover whether it
is caused by a leak, or by the refusal of the pump to work. He
can tell whether the pump is working by the sound of the check-
valve falling after each stroke, or by feeling the feed-pipe or
check-valve. A pump will not feed when the temperature of the
water is very high, unless it is specially adapted for pumping hot
water ; and if it refuses to work from this cause, the temperature
of the water should be reduced. A pump will not deliver water
if the proper valves are not opened, if its passages are choked, or
if its packing is defective. It would be necessary to examine the
pump at once, and endeavor to discover and remedy the difficulty.
If the water falls in the boiler on account of a leak, it can sometimes be temporarily repaired with a plug, or the pump can be
run faster, so as to keep up the water until stopping-time. If
this is not possible, the fire should be hauled, and the engine allowed to run as long as there is sufficient steam-pressure. In case
the engineer finds that the pump is not feeding, and he has a fair
supply of water in the boiler, he should at once examine the pump,
and endeavor to remedy the trouble without stopping the engine.
If he does not succeed, however, before the water falls below the
level of the lowest gauge-cock, he should haul the fire, and let the
engine run as long as the steam-pressure is sufficient. If he has
been called away from the boiler, and on his return finds that the
water is below the level of the lower gauge-cock, he should immediately ascertain the steam-pressure, and if it is rising rapidly he
should haul the fire at once. If the steam-pressure is about the
same as usual, he should examine the pump ; and if it is not de-
livering water, he should haul the fire. If the pump is feeding,
he may run it faster, watching the steam-gauge carefully. If
the pressure does not fall, he should stop the pump and haul the
fire. In any case the engine should not be stopped until the steam-

pressure is considerably reduced. The engineer should be very particular, on finding the water low, to examine the steam-gauge at once ; and if the pressure is unusually high, he should haul the fire without delay.

A boiler foams or primes either because it has insufficient steam room, or on account of dirt or grease in the boiler or the feed-water. The trouble is often experienced with new boilers, and disappears when they become clean. Priming is dangerous, if much water is carried over with the steam, as it is difficult to maintain the water-level constant, and the engine is liable to be broken by the water in the cylinders. If the trouble is caused by insufficient steam-room, it can sometimes be partially overcome by increasing the steam-pressure, and throttling it down to the ordinary working pressure in the cylinder ; but the only effectual way is to provide more steam-room. If the priming is due to dirt or grease in the boiler, the engineer should blow off frequently, and clean the boiler every few days. In blowing off, it is well to raise the water-level in the boiler by putting on a strong feed, and then blow down below the level that is ordinarily maintained. It is very often the case that the water-level is higher when the engine is running than it is when none of the steam is being used. The engineer should ascertain how much higher the water rises in such a case, so as to have a proper quantity of water when the engine is stopped. B.

Engine for a row-boat.—For an ordinary Whitehall row-boat, 18 feet long, to run at a speed of 8 miles per hour, the engine should have two cylinders, 2 in. diameter and 3 in. stroke ; tubular boiler, 24 to 28 inches in diameter, 4 feet high ; propeller, 22 to 24 inches in diameter, with 3 feet pitch.

Engines and Boilers, Small.—By the aid of the accompanying table, the effective horse-power (that available for useful work) of small engines can be approximately determined. The table is designed for non-condensing engines, with cylinders up to 6 inches in diameter, and for piston-speeds up to 400 feet a minute ; the connection of the engine with the boiler being supposed to be tolerably direct, the ports and pipes being of sufficient sizes, and the steam-valve closing when the piston has made three quarters of the stroke. As the table is designed for *average* conditions, it is evident that it will give results that are too large in some cases, and too small in others.

I.—To find the horse-power of an engine corresponding to a given diameter of cylinder, length of stroke, number of revolutions per minute, and pressure of steam in the boiler. (1) Multiply the length of stroke in inches by the number of revolutions per minute, and divide the product by 6. The result is the piston-speed in feet per minute. (2) Find the number in the table the nearest to the given steam-pressure and calculated piston-speed, and multiply it by 0.7854 times the square of the diameter of the piston in inches. *Example:* An engine has a cylinder 2 inches in diameter, and a length of stroke of 2 inches. It makes 400 revolutions a minute, and the boiler-pressure is 50 pounds per square inch. *Ans.:* Twice 400 is 800. 800 divided by 6 is 133⅓, the piston-speed in feet per minute. (3) The nearest piston-speed in table

Effective Horse-Power of an Engine with a Piston One Square Inch in Area, for different Steam-Pressures and Piston-Speeds.

Pressure.*	Horse-power corresponding to piston-speed (in feet per minute) of												
	10	20	30	40	50	60	70	80	90	100	200	300	400
10	.0005	.0010	.0015	.0020	.0025	.0030	.0035	.0040	.0045	.0050	.0099	.0149	.0199
15	.001	.003	.004	.005	.007	.008	.010	.011	.012	.014	.027	.041	.054
20	.002	.004	.007	.009	.011	.013	.016	.018	.020	.022	.045	.067	.089
25	.003	.006	.009	.012	.015	.019	.022	.025	.028	.031	.062	.093	.124
30	.004	.008	.012	.016	.020	.024	.028	.032	.036	.040	.079	.119	.158
35	.005	.010	.015	.019	.024	.029	.034	.039	.044	.048	.097	.145	.194
40	.006	.011	.017	.023	.029	.034	.040	.046	.051	.057	.114	.171	.228
45	.007	.013	.020	.026	.033	.039	.046	.053	.059	.066	.131	.197	.263
50	.0074	.015	.022	.030	.037	.045	.052	.060	.067	.074	.148	.223	.297
55	.008	.017	.025	.033	.042	.050	.058	.067	.075	.083	.166	.250	.333
60	.009	.018	.028	.037	.046	.055	.064	.073	.083	.092	.184	.275	.367
65	.010	.020	.030	.040	.050	.060	.070	.080	.090	.100	.201	.301	.402
70	.011	.022	.033	.044	.055	.065	.076	.087	.098	.109	.218	.327	.436
75	.012	.024	.035	.047	.059	.071	.083	.094	.106	.118	.236	.354	.472
80	.013	.025	.038	.051	.063	.076	.089	.101	.114	.127	.253	.380	.506
85	.0135	.027	.041	.054	.068	.081	.095	.108	.122	.135	.270	.406	.541
90	.014	.029	.043	.058	.072	.086	.101	.115	.129	.144	.288	.432	.575
95	.015	.031	.046	.061	.076	.092	.107	.122	.137	.153	.306	.458	.611
100	.016	.032	.048	.065	.081	.098	.113	.129	.145	.161	.323	.484	.645
105	.017	.034	.051	.068	.085	.102	.119	.136	.153	.170	.340	.510	.680
110	.018	.036	.054	.071	.089	.107	.125	.143	.161	.179	.357	.536	.715
115	.019	.038	.055	.074	.093	.112	.130	.149	.168	.187	.373	.560	.746
120	.0196	.039	.059	.076	.098	.118	.137	.157	.177	.196	.392	.588	.785
125	.020	.041	.061	.082	.102	.123	.143	.164	.184	.205	.410	.614	.819
130	.021	.043	.064	.085	.107	.128	.149	.171	.192	.213	.427	.640	.854
135	.022	.044	.067	.089	.111	.133	.156	.178	.200	.222	.445	.667	.889
140	.023	.046	.069	.092	.115	.139	.162	.185	.208	.231	.462	.693	.924
145	.024	.048	.072	.096	.120	.144	.168	.192	.216	.240	.479	.719	.958
150	.025	.050	.074	.099	.124	.149	.174	.199	.223	.248	.496	.745	.993

* In boiler, by gauge.

is 130 feet, and the number in table corresponding to speed of 100 and pressure of 50, is 0.074 ; the number for speed of 30 and same pressure is 0.022 ; required number is sum of 0.074 and 0.022, or 0.096, corresponding to speed of 130 and pressure of 50. The product of 4 and 0.7854 is 3.1416, and the product of 0.096 and 3.1416, or the required horse-power, is 0.3+.

II.—Diameter in inches of cylinder required for a given horse-power, piston-speed, and boiler-pressure. (1) Find in the table the number nearest to the given piston-speed and steam-pressure. (2) Multiply the number obtained in (1) by 0.7854. (3) Divide the given horse-power by the quantity obtained in (2). (4) Take the square root of the quantity obtained in (3).

Example.—What should be the diameter of cylinder of an engine developing 2 horse-power, with a piston-speed of 150 feet a minute, and a boiler-pressure of 100 pounds per square inch ? (1). The number from the table is the sum of 0.161 and 0.081, or 0.242. (2) The product of 0.242 and 0.7854 is 0.1900668. (3) The quotient of 2 divided by 0.1900668, is about 10.5226. (4) The square root of 10.5226 is 3.24+, or about 3¼ inches, the required diameter of cylinder.

III.—The number of pounds of water required to be evaporated per hour for each horse-power exerted, and for various boiler-pressures, may be approximately estimated from the accompanying table.

Pressure of steam in boiler by gauge.	Pounds of water per effective horse-power per hour.	Pressure of steam in boiler by gauge.	Pounds of water per effective horse-power per hour,
10	118	60	75
15	111	70	71
20	105	80	68
25	100	90	65
30	93	100	63
40	84	120	61
50	79	150	58

It is convenient, in calculations of the amount of water evaporated at various pressures and from various temperatures of feed, to reduce them to a common standard, namely, the equivalent amounts that would be changed into steam of atmospheric pressure, if the temperature of feed was 212° Fahrenheit ; or, as it is commonly called, to evaporation "from and at 212." Two tables are appended, for the purpose of facilitating this reduction. The second table is taken from Professor Rankine's "Treatise on the Steam-Engine."

Pressure and Temperature of Steam.

Pressure by gauge.	Temperature Fahrenheit.	Pressure by gauge.	Temperature Fahrenheit.
0	212°	60	307°
10	239°	70	316°
20	250°	80	324°
30	274°	90	331°
40	287°	100	338°
50	298°	110	344°

Pressure by gauge.	Temperature Fahrenheit.	Pressure by gauge.	Temperature Fahrenheit.
120	350°	170	375°
130	356°	180	379°
140	361°	190	384°
150	366°	200	388°
160	370°		

Factors of Evaporation.

Temperature of the steam.	Temperature of the feed-water.										
	32°	50°	68°	86°	104°	122°	140°	158°	176°	194°	212°
212°....	1.19	1.17	1.15	1.13	1.11	1.10	1.08	1.06	1.04	1.02	1.00
230°....	1.20	1.18	1.16	1.14	1.12	1.10	1.08	1.06	1.04	1.02	1.01
248°....	1.20	1.18	1.16	1.14	1.13	1.11	1.09	1.07	1.05	1.03	1.01
266°....	1.21	1.19	1.17	1.15	1.13	1.11	1.09	1.07	1.06	1.04	1.02
284°....	1.21	1.20	1.18	1.16	1.14	1.12	1.10	1.08	1.06	1.04	1.02
302°....	1.22	1.20	1.18	1.16	1.14	1.12	1.11	1.09	1.07	1.05	1.03
320°....	1.22	1.21	1.19	1.17	1.15	1.13	1.11	1.09	1.07	1.05	1.03
338°....	1.23	1.21	1.19	1.17	1.15	1.14	1.12	1.10	1.08	1.06	1.04
356°....	1.23	1.22	1.20	1.18	1.16	1.14	1.12	1.10	1.08	1.06	1.04
374°....	1.24	1.22	1.20	1.18	1.17	1.15	1.13	1.11	1.09	1.07	1.05
392°....	1.24	1.23	1.21	1.19	1.17	1.15	1.13	1.11	1.09	1.07	1.06
410°....	1.25	1.23	1.22	1.20	1.18	1.16	1.14	1.12	1.10	1.08	1.06

To illustrate the use of the tables, suppose an engine of 6 horse-power is supplied with steam at a pressure of 50 pounds per square inch, and that the temperature of the feed-water supplied to the boiler is 160°. It is required to find how much water must be evaporated per hour "from and at 212°" for the engine. The temperature of steam having a pressure of 50 pounds is 298°. In the table of "Factors of Evaporation," the factor corresponding to a steam temperature of 302° and a feed temperature of 158° (which are the numbers in the table nearest to the given ones), is 1.09. Now this engine requires an evaporation of 6 times 79, or 474 pounds of water per hour, at a pressure of 50 pounds, or an equivalent evaporation "from and at 212°" of 1.09 times 474, which is equal to 516.66 pounds.

IV.—To find the proportions suitable for a boiler which is to have a given evaporation. (a) *To ascertain the grate-surface in square feet :* Divide the number of pounds of water to be evaporated per hour, from and at 212°, by 75, for cylinder boilers ; by 77, for flue boilers ; by 78, for tubular boilers ; by 80, for locomotive and vertical boilers.

Example.—Suppose that a cylinder boiler is to be proportioned for an evaporation of 500 lbs. of water per hour, at a pressure of 75 lbs., the temperature of the feed-water being 80°. The equivalent evaporation will be 1.17 times 500, or 585 lbs., and the grate-surface 585 divided by 75, or $7\frac{8}{10}$ square feet.

(b) *To ascertain the heating surface in square feet :* Multiply the grate-surface by 11, for cylinder boilers ; by 17, for flue boilers ; by 30, for tubular, locomotive, and vertical boilers. (c) *To ascertain the cross-section of flues or tubes in square feet :* Multiply the grate-surface by 0.134. This is an average value for good practice, and it can be varied between the limits of 0.125 and 0.143, as may be most convenient. (*d*) *To ascertain the length of boiler :* Cylinder boilers should be from 10 to 12 times the diameter ; flue boilers from 5 to 6 times the diameter ; tubular boilers, and the shells of locomotive and vertical boilers, from 3 to $3\frac{1}{2}$ times the diameter. There is very great variation from these figures in practice ; but the numbers given above represent the most general limits, so far as they can conveniently be classified.

Example.—What are the dimensions of a tubular boiler for an engine that is to develop $4\frac{1}{2}$ horse-power, with a steam-pressure of 100 lbs., the temperature of the feed-water being 160° ?

The equivalent evaporation required per horse-power per hour is 1.1 times 63, or $69\frac{3}{10}$ lbs. The total equivalent evaporation is $4\frac{1}{2}$ times $69\frac{3}{10}$, or about 312 lbs. Hence the grate-surface, being the quotient arising from dividing 312 by 78, is 4 square feet. The heating surface is 30 times 4, or 120 square feet. The cross-section of the tubes should be about 0.536 square feet (4 times 0.134), or it should vary between the limits of 0.5 (4 times 0.125) and 0.572 (4 times 0.143) square feet.

While the rules relating to engines given above are generally only applicable for cases within the limits mentioned at the beginning of this article, those for the proportions of boilers give safe average values for the majority of cases that are met with in practice. **B.**

ENGINES, Derangements of.—These are hot bearings, loose keys, and leaky joints. If a bearing heats continually, when properly adjusted and well lubricated, it is too small. Sometimes bearings heat on account of dirt or grit, because they are set up too tightly, or are out of line. A hot bearing can often be cooled without stopping the engine, by mixing sulphur or black-lead with the oil, or by turning on a stream of water from a hose. If a joint blows out it can sometimes be wedged, so that the engine can be run until stopping time. An engineer should exercise all his ingenuity to overcome a difficulty without stopping the engine, except in cases where it would be dangerous to continue to run. If keys or bolts become loose, it will generally be indicated by a thump in the engine. To prevent the freezing of pipes and connections in exposed situations, they should either be thoroughly drained, or the water should be kept circulating in them. **B.**

EXHAUST STEAM.—This should not be discharged into a brick chimney. It is liable to disintegrate the mortar and destroy the chimney.

STEAM-ENGINE GOVERNORS.—The ordinary pendulum governor consists of a vertical spindle, which is made to revolve by suitable mechanism, and carries, on opposite sides, a pair of arms, to which heavy weights are attached, forming revolving pendulums, which vary their positions at different speeds. The simplest form of construction is shown in Fig. 1, A B being the revolving spindle, E and D the weights, secured to the spindle by rods jointed at G. Several positions of the balls are shown, corresponding to different speeds of rotation. In any of these positions, the vertical distance, as G F, of the point of suspension, G, above the centres of the balls, is called the *height of the governor*, and it can be found for any case by dividing 32,508 by the square of the number of revolutions per minute. For instance, if a governor makes 100 revolutions per minute, running without friction or other resistance, the vertical distance of the centres of the balls below the point of suspension would be 32,508 divided by 10,000 (the square of 100), or about 3½ inches. A table is added, showing the heights corresponding to various speeds.

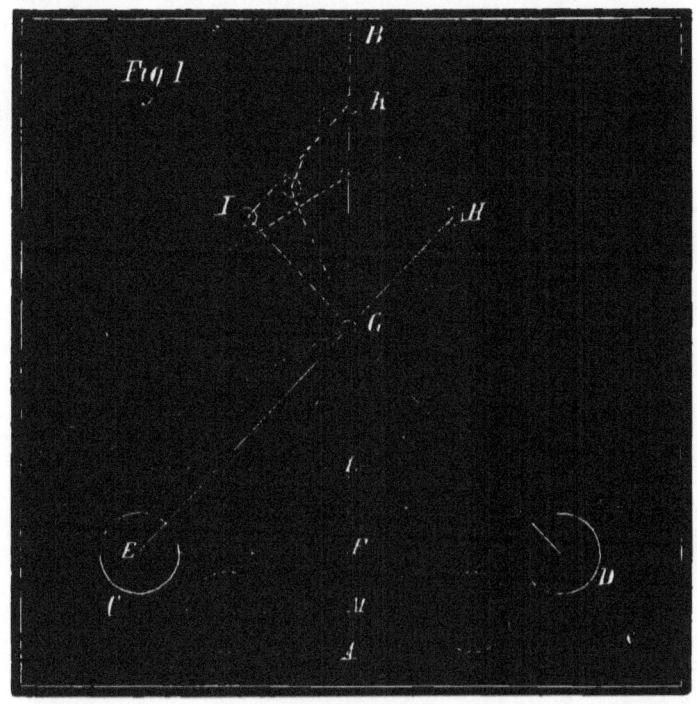

Table.

Revolutions per minute.	Height in inches.	Revolutions per minute.	Height in inches.
10	352.08	275	0.4646
20	88.02	300	0.3912
30	39.12	350	0.2873
40	22.01	400	0.2201
50	14.08	450	0.1739
60	9.78	500	0.1408
70	7.184	550	0.1164
80	5.501	600	0.0978
90	4.347	650	0.08333
100	3.521	700	0.07184
125	2.253	750	0.06259
150	1.564	800	0.05501
175	1.150	850	0.04873
200	0.8802	900	0.04347
225	0.6955	950	0.03901
250	0.5633	1000	0.03521

In practice, the pendulum governor is generally constructed somewhat as represented in Fig. 2, being connected to the controlling mechanism by short levers, so that a slight change in the position of the balls will move the regulator considerably. In estimating the height of the balls of such a governor, it is to be measured from E, where the centre lines of the arms produced cut the centre of the spindle.

When a governor acts on the controlling mechanism of an engine, it encounters some resistance. There is also some friction of the moving parts, and a weight is sometimes added, either sliding on the spindle or connected to the spindle by a lever, in order to make the governor more sensitive. All these things influence the height of the governor. In a well-made instrument the friction is insignificant, and need not be regarded, but allowances must be made for the resistance, and the weight, if any is attached. Find how many pounds of force are required to move the controlling mechanism of the engine, and find the weight

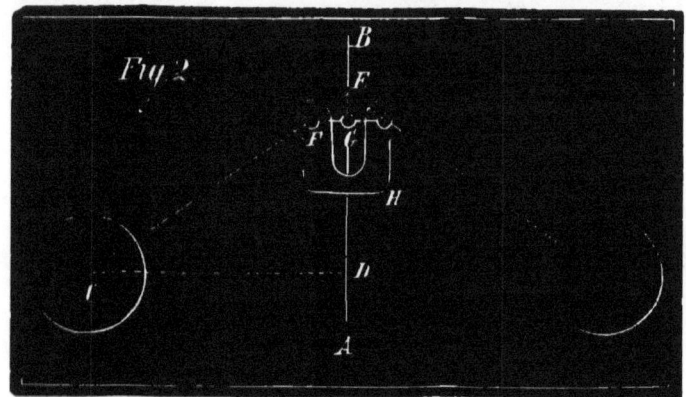

of the balls and of the attached weight, in pounds. Next determine how far the controlling mechanism is moved, and the attached weight raised or lowered, for a given change in the height of the governor-balls. Divide the distance moved by the resistance by the change in height of the balls in the same time, and multiply the quotient by the measure of the resistance in pounds; divide also the vertical distance moved by the attached weight for a given change in the height of the balls, and multiply the quotient by this weight. Take the sum of these two products and the weight of the governor-balls, and divide by the weight of the governor-balls; multiply the quotient by the height for a governor working freely, taken from the table above : the quotient is the corrected height of the governor-balls.

Example.—The two balls of a governor weigh 20 lbs.; the resistance of the mechanism is 10 lbs., and it moves 4 in. while the height of the balls changes $\frac{1}{4}$ inch. The attached weight is 8 lbs., and it moves 2 inches vertically, while the height of the balls changes $\frac{1}{4}$ in. What is the height of balls for a speed of 200 revolutions a minute? Multiplying the quotient of 1 divided by $\frac{1}{4}$ (4) by 10, the product is 40 ; multiplying the quotient of 2 divided by ¼ (8) by 8, the product is 64 ; dividing the sum of 40, 64, and 20 (124) by 20, the quotient is 6.2 ; multiplying 6.2 by 0.8802 (the height for a free governor making 200 revolutions a minute), the product, *the corrected height* of the governor-balls, is about 5¼ inches.

In designing a governor, it is well to fix upon some range of speeds between which it shall control the engine, and make the balls heavy enough to effect this. The proper weight for the balls can be found approximately, as below:

(1) Divide the distance through which the resistance moves by the change in height of the governor-balls in the same time, and multiply the quotient by the resistance ; divide the vertical distance through which the attached weight moves by the vertical distance moved by the balls in the same time, and multiply the quotient by this attached weight ; add together these two products, and divide the sum by 2.

(2) Subtract the mean speed of the governor from the greatest speed it is to have, and divide the difference by the mean speed ; divide the quantity obtained in (1) by this quotient : the result will be the weight of the two balls.

Example.—Suppose the resistance and attached weight are the same as in the preceding example, and that the speed of the governor is to vary between 200 and 300 revolutions a minute, in controlling the speed of the engine. What should be the weight of the balls?

(1) The sum of 40 and 64 (the corrected resistance and attached weight) is 104 ; ½ of 104 is 52.

(2) The difference between 300 and 200 (100), divided by 200, is 0.5 ; the quotient of 52 divided by 0.5, or *the weight of the balls*, is 104 lbs., so that each ball must weigh 52 lbs.   B.

HORSE-POWER, Different kinds of.—In making an estimate or measure of the effect of any piece of mechanism that is used to overcome resistance, it is necessary to have a unit of reference. In whatever manner the resistance is overcome, if it can be measured it can be converted into the amount of work that must be expended to raise a weight through a distance, since by suitable arrangements the mechanism can be put in motion and made to overcome resistance by allowing the weight to fall. This gives a simple mode of estimating the work, by assuming that a unit of work is the amount required to raise 1 pound a distance of 1 foot vertically. To illustrate, suppose that a cut is being taken from a 6-in. shaft in a lathe, and that the resistance to the motion of the cutting-tool is 200 lbs.; how many units of work are performed each time the shaft makes a revolution?

In each revolution of the shaft, the tool makes a cut 6 times 3.1416, or 18.8496 ins., or 1.5708 ft. in length, and the work is the same as would be required to raise a weight of 200 lbs. through a vertical distance of 1.5708 ft., or it is 314.16 units.

Now, if 33,000 units of work are performed in a minute, they constitute a unit of power, known as a *horse-power*—and conversely, a horse-power can be defined as the power required to raise 33,000 lbs. 1 ft. high, or do 33,000 units of work in a minute. Again, a horse-power may be defined as the power required to perform 550 units of work in a second, or 1,980,000 units in an hour. To apply the principle to the example given above, suppose the shaft makes 20 revolutions a minute, how many horse-power are required to drive the tool? In this case, the units of work performed per minute would be 20 times 314.16, or 6283.2, and the horse-power would be $\frac{1}{33,000}$ of 6283.2, or about $\frac{19}{100}$ of a horse-power. It will be seen, from this e ample, that if the resistance to motion in pounds and the speed of motion in feet per minute can be measured, it is only necessary to multiply them together, and divide by 33,000, in order to obtain the horse-power. The steam-engine is a machine that is commonly rated as being of a certain horse-power, but the term horse-power, as thus used, does not always have the same meaning. In fact, there are four kinds of horse-power by which an engine may be rated :

1. *Gross* or *indicated* horse-power.
2. *Net* or *effective*   "
3. *Total*   "   .
4. *Nominal*   "

1. The *gross* or *indicated* horse-power of an engine is the power calculated by assuming the resistance to be that due to the mean effective pressure of the steam on the piston, as shown by the indicator. Thus, suppose this pressure is 2500 lbs., and the piston moves 400 ft. a minute, the gross horse-power is 400 times 2500 divided by 33,000, or 30.3.

2. The *net* or *effective* horse-power of an engine is computed from the useful resistance overcome. If, in the preceding example, the pressure on the piston, after deducting that required to overcome the friction of the engine, is 2200 lbs., the effective horse-power is 2200 times 400, divided by 33,000, or 26.7.

The net horse-power is the proper kind to be specified by a steam-user when he is buying an engine.

3. The *total* horse-power of an engine is computed from the total pressure of the piston above a vacuum. If, in the example given in Case 1, the total pressure on the piston is 4200 lbs., the total horse-power is 400 times 4200, divided by 33,000, or 50.9.

Total horse-power is only used in comparisons of the results of experiments.

4. The *nominal* horse-power of an engine has no meaning in particular—that is to say, there are a number of different rules by which it may be computed. Thus, there is the admiralty rule for marine engines, Mr. Bourne's rule for condensing engines, Mr. Bourne's rule for non-condensing engines, James Watt's rule ; and numerous engine-builders have private rules of their own. For instance, *A* says, "I will make an engine with a cylinder 10 ins. in diameter, and a stroke of 15 ins., and I will call it 8 horse-power, nominal."

B, who builds an engine of the same size, and wants to make purchasers think they are getting more for their money, says, " I will call my engine 16 horse-power, nominal." The man who goes to buy a steam-engine of either of these parties may very properly say to them, " How much will you charge me for an engine guaranteed to be of so many horse-power, actual?" B.

INDICATOR, The steam-engine.—The construction is shown in Figs. 1 and 2, Fig. 1 being an elevation, and Fig. 2 a section.

The indicator is a recording steam-gauge, very accurately made, for determining the pressure acting on the piston of an engine, at every point of the stroke. It is connected to the cylinder, close to one end, and when the cock, seen in Fig. 1, is opened, the steam presses on a small piston, shown in Fig. 2. A stiff spiral spring above this piston is compressed by the pressure of the steam. The piston-rod, it will be seen, is connected to a lever, and this, in turn, with a link and another lever, a pencil or marking-point being placed in a hole in the link. There is a cylindrical barrel to the left on which a piece of paper can be placed, being held by two clip springs. This barrel can be made to revolve by pulling a string wound round the bottom, and it has within it a coiled spring, which makes it turn back again

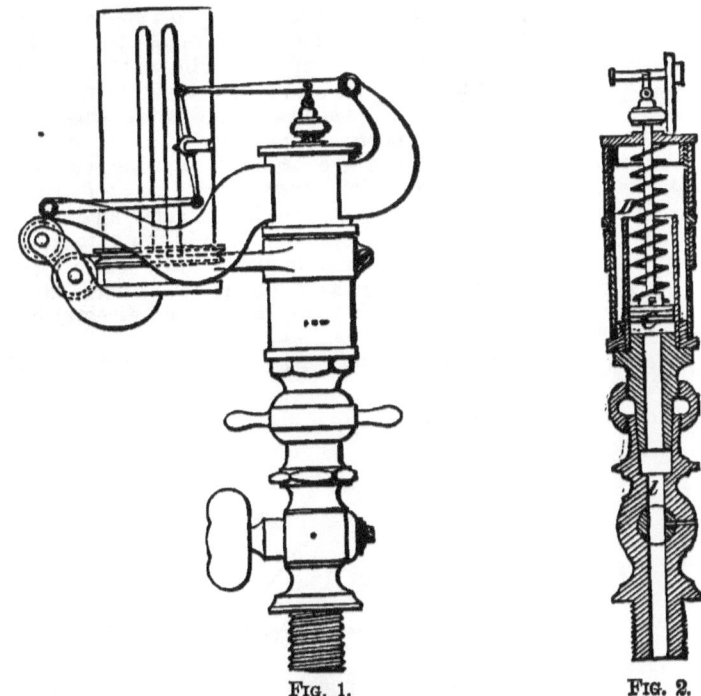

FIG. 1.   FIG. 2.

THE INDICATOR.

when the tension of the string is relaxed. Now, suppose that the indicator is attached to the cylinder of an engine, and the cord is fastened to some moving part, so that when the engine makes a stroke it causes the barrel carrying the paper to revolve, and on the return stroke the coiled spring in this paper barrel makes it turn back to its original position. Meanwhile the steam in the cylinder is pressing on the piston of the indicator, forcing it up a distance corresponding to the pressure, so that if the pencil is allowed to touch the paper, it will trace out a line which represents the pressure, and this line is called an *indicator diagram*. Such a diagram is shown in Fig. 7. The atmospheric line, *C D*, is traced when the cock is closed and there is no pressure on the piston. At *b*, the stroke of the engine commences, and when the piston has gone about half way to the other end of the cylinder, the steam is cut off, as shown at *c*, and the pressure begins to fall as the steam expands. Near the end of the stroke, the exhaust-valve opens, as shown at *d*, and the pressure falls more rapidly. When the engine makes the return stroke, there is only the back pressure, until near the end, when the exhaust-valve closes, shown at *f*, and the steam being compressed, the pressure rises. Just before the end of the stroke, at *a*, on the diagram, steam is admitted, and the pressure rises suddenly.

I. *How to attach the indicator to the cylinder of an engine.*

Drill a hole in the cylinder, in the head, or close to the end, and tap it out for a half-inch iron nipple. The indicator-cock must be connected to this, using an elbow, if necessary, and then the indicator can be attached at pleasure. In drilling this hole, do not make it close to the ports. Sometimes the connections from the two ends of a cylinder are brought together, and one indicator is made to answer for both ends, a cock being fitted in each pipe, so that either can be opened to the indicator, as desired. In such a case, the holes in the cylinder should be larger, for three-quarter inch pipe, at least, so as to prevent any loss of pressure. It is obvious, however, that as the indicator is used to obtain the pressure in a cylinder, the more closely and directly it is connected, the better.

II. *How to make the paper barrel have a motion coincident with that of the piston.*

a. Reducing-wheel.—Fig. 3.

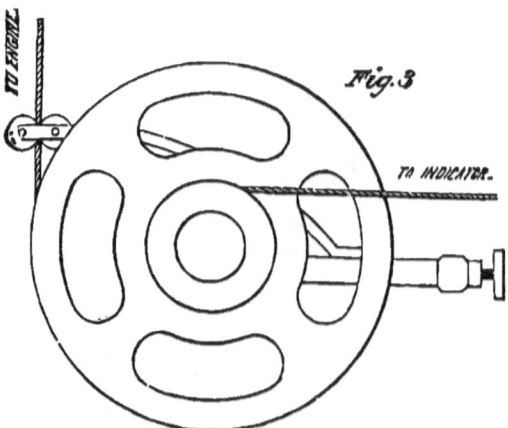

This is attached to some part of the engine-frame, and the cord marked " to engine" is made fast to the cross-head, being carried over a pulley, if necessary, so that it is parallel to the guides. The other cord is fastened to the cord wound round the paper barrel of the indicator. The two wheels bear the same proportion to each other as the stroke of the engine does to the desired range of motion for the paper barrel, the latter being usually from 4 to

5 inches. There is a coiled spring in the reducing-wheel, which makes it turn back on the return movement of the cross-head. By having different sized wheels to carry the cord leading to the indicator, this arrangement can be adapted to engines with different strokes.

b. *Swinging-board, with slot.*—Fig. 4.

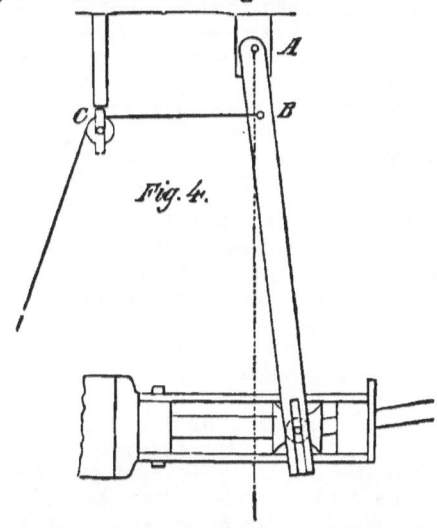

Fig. 4.

Mark a point, A, at some convenient distance from the cross-head, and on a line perpendicular to the guides, at the centre of the stroke. Attach a board so that it can swing freely around this point; cut a slot in the other end, for a pin connected to the cross-head. Then, as the cross-head moves, it will make the board swing to and fro. At some point B, of the board, which

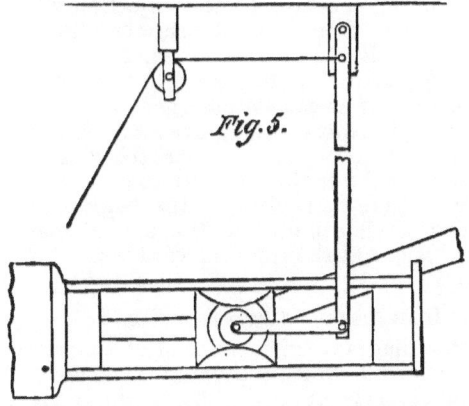

Fig. 5.

has the proper movement for the paper cylinder, attach a cord or wire, and carry it over a pulley, C, adjusted at such a height that the part of the cord, $B\ C$, is parallel to the guides when the engine is at half stroke. The cord can then be brought down and attached to the cord of the paper barrel.

c. Swinging-board, with link.—Fig. 5.

Sometimes it is not practicable to attach the board directly to the cross-head by a pin, and it is more convenient to use a link connection, the arrangement of which will be evident from the figure.

It is easy to see that a number of arrangements could be devised on the general principle of the swinging-board. Sometimes it is attached to the guides by a standard, and sometimes one end is connected to the cross-head and the other to the indicator, the point around which the board swings being between, at distances from the two ends proportional to the stroke of the engine and the movement of the paper barrel. Whatever the special arrangement, attention should be given to these two points :

1st. To have the board perpendicular to the guides when the engine is at half stroke.

2d. To lead the cord off in a direction parallel to the guides.

III. *How to take an indicator diagram.*

If a cord is used for the motion, it should have a slide on it, so that it can be adjusted, and it should have a hook so that it can be attached to the cord of the paper barrel, and detached at pleasure. Fine wire is better than cord, as it is quite flexible, and does not stretch so readily. If several cards, taken at intervals, are of the same length, the connection is all right. Having got the motion properly adjusted, turn the cock of the indicator so that it will blow through, having first put a piece of paper on the paper barrel. Then turn the cock so as to let steam into the indicator, and press the pencil lightly against the paper ; draw it back as soon as the card is traced, shut the indicator-cock, and apply the pencil again, to trace the atmospheric line—be particular not to trace the atmospheric line until the diagram is taken ; then unhook the cord, remove the diagram from the paper cylinder, and mark on it the pressure of steam by gauge, the revolutions per minute, the height of the barometer, and the temperature of the engine-room—if the proper instruments are available—and the vacuum by gauge, the temperature of the hot-well, and the temperature of the injection-water, if the diagram has been taken from a condensing-engine. All the dimensions of the engine should also be noted, for future reference, together with such particulars in regard to the dimensions and performance of the boiler as can be obtained.

A counter should always be employed to determine the number of revolutions per minute, at the time the diagram is taken. It can be connected to the indicator motion, or to some moving part of the engine. Take its reading at the beginning of a minute, and at the end, having indicated the engine meanwhile. This gives the revolutions at the time the card was taken, with considerable accuracy.

IV. *How to draw the true diagram.*—Fig. 6.

The indicator diagram from one end of the cylinder, it is evident, only shows what takes place on the side of the piston on which the indicator is applied—but at the same time there is some back pressure on the other side, opposing the motion of the piston. To get the actual diagram, therefore, it is necessary to take diagrams from both ends of the cylinder simultaneously, and then combine the parts of each that were traced at the same time. Thus such a figure as is shown in Fig. 6 is obtained. On one side of the piston, the line *a b c f* is traced, and *d c e* is traced at

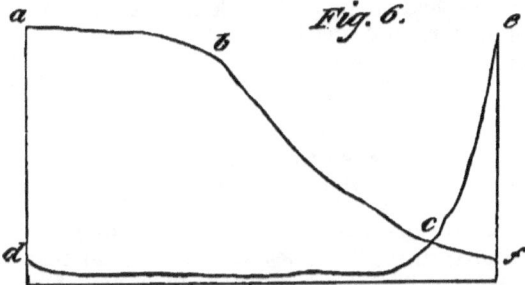

Fig. 6.

the same time on the other, and the figure so obtained is the true diagram representing the distribution of the pressure. In using the diagram as ordinarily taken, therefore, the actual effective pressure is not obtained ; but there is no error in practice, since the inaccuracies of the diagrams from the two sides balance each other. It is a great mistake, however, to take diagrams from one end of a cylinder only, and assume that those from the other would be similar. Quite often there are serious differences, and both ends of a cylinder should always be indicated, if possible

V. *How to ascertain the mean effective pressure from a diagram.* —Fig. 7.

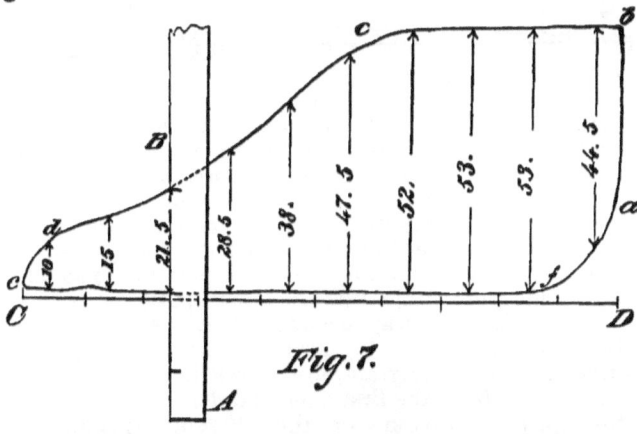

Fig. 7.

a. 1st *method.*—Draw perpendiculars to the atmospheric line, from the extremities of the diagram, thus determining its length, *C D;* divide the line, *C D*, into 10 equal parts, and midway between each of the divisions erect a perpendicular to *C D*, drawing it between the upper and lower boundaries of the diagram ; measure the length of each of these lines on the scale of the indicator-spring, add the measurements together, and divide the sum by 10. In the figure, the pressure, or length of the line on the scale, is shown at each perpendicular. The sum of these is 363, so that the mean effective pressure is 36.3 lbs. per square inch.

b. 2d method.—Draw perpendiculars between the ten divisions of *C D,* as explained above. Then take a strip of paper, apply it to the first perpendicular, and mark the length ; apply it to the next perpendicular, and mark its length, next to the first; so continue applying it to each, and when the last perpendicular has been measured, the distance between the first and last marks will be the sum of all the lengths. The strip of paper, *A B,* is shown in the figure as applied to the third perpendicular. Measure the length of the paper between the extreme marks, in inches, multiply it by the scale of the indicator-spring, and divide by 10. Suppose, in the present case, that the length of the paper is found to be 12.1 inches, and that each inch represents 30 lbs. on the scale of the indicator-spring ; 30 times 12.1 is 363, so that the mean effective pressure is 36.3 lbs. per square inch, as before.

c. Positive and negative pressure.—Fig. 8.

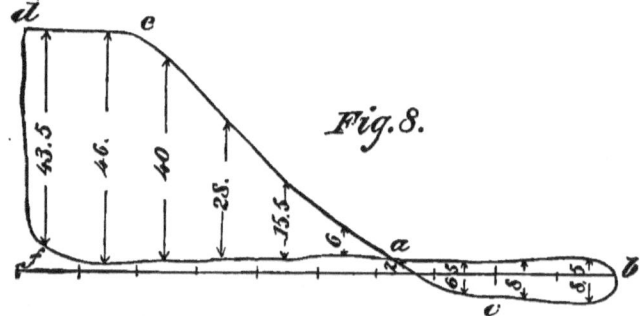

When the steam is cut off very early in the stroke, and the valves and piston are tight, a diagram is sometimes drawn like that in the figure, in which the back pressure is greater than the forward pressure for a portion of the stroke, and the pressure determined from the portion of the diagram *a c b* must be subtracted from the pressure due to the portion *d e a f.* In such a case, the method of determining the mean pressure is as follows : Divide the atmospheric line into 10 equal parts, as before, and draw perpendiculars midway between them ; add together the perpendiculars (measured in the scale of the indicator-spring) in the positive part of the diagram, also those in the negative part ; subtract the latter from the first sum, and divide the difference by 10. In the figure, the pressures at the different perpendiculars are given. The sum of the positive pressures is 179, of the negative pressures 25, and the difference is 154; so that the mean effective pressure is 15.4 lbs. per square inch.

VI. *How to find the indicated horse-power of a steam-engine.*

Having determined the mean effective pressure from a diagram, by one of the methods explained above, multiply this pressure by the product of the stroke in feet, the square of the diameter of the cylinder in inches, the number of revolutions per minute, and 0.0000476.

Example.—Suppose the mean effective pressure is 50 lbs. per square inch, the diameter of the cylinder 15 inches, the length of stroke 2 feet, and the number of revolutions per minute 80. Then the horse-power is the product of 50,225 (the square of 15), 80, and 0.0000476, or 85.68.

VII. *How to construct the theoretical diagram.*—Fig. 9.

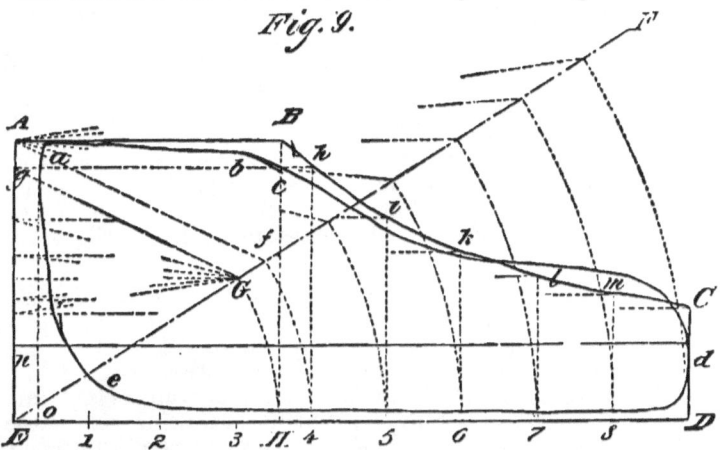

Fig. 9.

This is the diagram that would be taken if the steam acted in the cylinder with the pressure at the beginning of the stroke until the point of cut-off, and that then the admission ceased instantaneously, and the steam expanded, in accordance with Mariotte's law, to the end of the stroke, when the exhaust-valve opened, and the steam was immediately condensed, creating a perfect vacuum in the cylinder for the return stroke. Such a diagram is represented by *A B C D E*, this being the theoretical diagram for the actual diagram, *a b c d e*. The following is the method of laying it down: Draw a line, *E D*, at a distance below the atmospheric line, *n d*, equal to the pressure of the atmosphere (14.7 lbs. per square inch on an average), on the scale of the diagram ; mark on *E D* the length, *o D*, of the actual diagram ; then find the total volume of the clearance spaces at the end of the cylinder from which the diagram was taken, and make *o E* bear the same relation to *o D* as this volume of clearance has to the total volume swept through by the piston per stroke. · To make this plain, suppose that in a cylinder having a diameter of 24 inches and a stroke of 3 feet, it is found that the volume of the clearance spaces at one end of the cylinder is 900 cubic inches. The volume swept through by the piston per stroke is the product of 0.7854, 24 squared, and 36 or 17,286 cubic inches, so that the clearance is about $\frac{52}{1000}$ of the piston displacement, and *o E* must be made $\frac{52}{1000}$ as long as *o D*. Thus, if *o D* is 5 inches, *o E* must be $\frac{26}{100}$ of an inch. Having determined the point *E*, make *E A* perpendicular to *E D*, and draw a line, *A B*, parallel to *E D*, at such a height that it represents the initial pressure of the steam. Through *c*, on the actual diagram, where the steam is cut off, draw a perpendicular, $B H$, to $E D$. Divide $E D$ into any number of equal parts, and erect perpendiculars at the points of division that are beyond the point of cut-off. From E draw any diagonal line, $E F$, and from E as a centre, with a radius $E H$, draw an arc cutting $E F$ in the point G. From the same centre, and with radii equal to $E 4$, $E 5$, etc., draw arcs cutting $E F$. The arc drawn with a radius $E 4$, cuts $E F$ in the point f. Draw the line $f A$, and from G draw a line, $G g$, parallel to $f A$. From g draw a line, $g.h$, parallel to $f A$, and the point h, in which it cuts the perpendicular drawn through 4, is a point of the curve of expansion. The construction of the points on the other perpendiculars is precisely similar, and is indicated in the figure. Having determined a sufficient number of points, draw through them the curve of expansion, $B h i k l m C$, and the theoretical diagram will be completed.

This kind of diagram is useful for comparing the merits of different engines, since it is evident that, other things being equal, the engine whose actual diagram most closely approaches the theoretical is the best. The mean pressure, as shown by such a diagram, can be determined by one of the methods already explained, for comparison with the pressure given by the actual diagram.

VIII. *How to take care of the indicator.*—Always oil the cylinder and all moving parts before applying the instrument to an engine. Never use any thing but the finest grade of oil, such as that specially prepared for sewing-machines or clocks. After taking one or two diagrams, remove the indicator, and examine its steam-cylinder. If any grit has entered, wipe it out, using soft cotton-waste on the end of a white-pine stick. It is important to attend to this on indicating an engine of which the condition is not known, and if it is found that dirt or grit is forced into the indicator, it should be cleaned at frequent intervals during the experiments. As soon as the experiments have been concluded, remove the indicator, and when it has cooled sufficiently, wipe it out and apply oil in the steam-cylinder, and to all the moving parts. Be careful to dry the spring thoroughly, and cover it with oil. Never put any hard substance into the steam-cylinder, but use a white-pine stick and soft cotton-waste. After use, take out the plug of the indicator-cock, clean it and the seat, apply oil, and on replacing it adjust it so that it moves freely and does not leak.

It might be supposed that minute directions of this kind were superfluous, and that any one owning an indicator would see the necessity of using such a delicate instrument with great care. The generality of the indicators in common use, however, impress the observer with the idea that all the directions recited above have been studiously neglected   B.

SLIDE-VALVES. Setting.—The methods of adjusting the lap and travel of slide-valves, and the position of the eccentric, given below, are taken, with some slight modifications, from the work of Dr. Zeuner, on Slide-Valve Gearing. It is believed that the simplicity of the construction will be appreciated by the reader.

I. *Area of ports.*—To find the proper area of port for an engine of a given piston-speed, multiply the area of the piston,

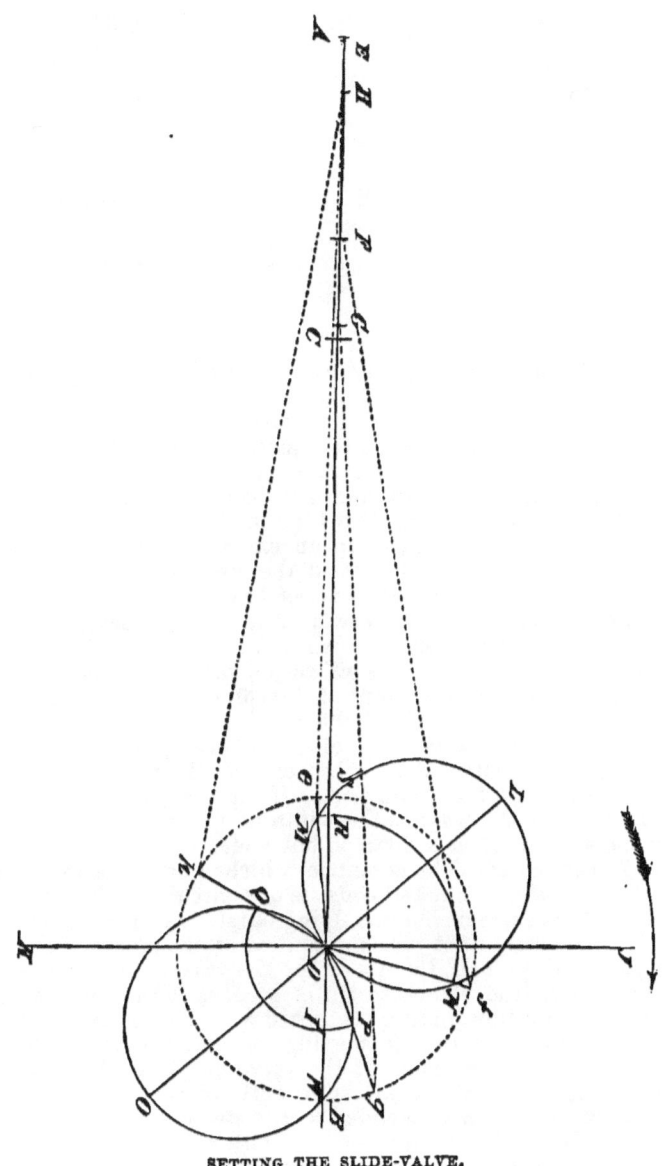

SETTING THE SLIDE-VALVE.

in square inches, by the number nearest to the given piston-speed
in the table on the next page.

Speed of piston, in feet, per minute.	Number by which area of piston is to be multiplied.
100	0.02
200	0.04
300	0.06
400	0.07
500	0.09
600	0.1
700	0.12
800	0.14
900	0.15
1000	0.17
1100	0.19
1200	0.2
1300	0.22
1400	0.24
1500	0.25

II. *Lap, lead, and travel of valve.*—The amount of opening given by the valve for the admission of steam or its exhaust, at the commencement or termination, respectively, of the stroke of an engine, is called the lead, either steam or exhaust, as the case may be. If the face of the valve is wider than the port, the excess of width is called lap, and may be either steam or exhaust lap. Steam-lap is an excess of width on the outer extremities of the valve-faces, and exhaust-lap an excess on the inner faces. The effect of steam lap is to cut off the steam at an earlier point of the stroke, and exhaust-lap causes the exhaust to open later and close earlier than it otherwise would. It must be obvious that the action of the exhaust would be very much deranged if an attempt was made to cut off very short, and it is found, in practice, that the limiting point of cut-off with the simple slide-valve is at about two thirds of the stroke.

The travel of the valve is the distance between its two extreme positions. For a valve without lap or lead, the travel is equal to twice the width of the steam-port. If lap is added, the travel of the valve is equal to twice the width of the port, increased by twice the amount of steam-lap on one end.

III. *The eccentric.*—The eccentric, which moves the valve, is a substitute for the crank, and consists of a circular disk secured to the shaft, the centre of the disk lying outside of the centre of the shaft. The distance between the centres of the eccentric and the shaft is equal to half the travel of the valve. If a valve has neither lap nor lead, the eccentric is secured to the shaft in such a position that a line joining its centre with the centre of a shaft is perpendicular to a line connecting the centre of the shaft and the centre of the crank-pin. If the valve has lead, the eccentric must be turned on the shaft sufficiently to secure the desired amount, and if lap is also added to the steam side, the eccentric must be advanced still further. In either of these cases, the amount the eccentric is moved forward is termed the angular advance of the eccentric, being the angle made by a line joining the centres of the eccentric and shaft with a line drawn through the centre of the shaft perpendicular to the line of centres of the crank-pin and shaft. The points in which these two perpendicular lines intersect the shaft should be plainly marked by a centre-punch or chisel, for convenience in adjusting the eccentric.

IV. *Proportions of valve and seat.*—The bridge of the valve-seat should generally have a width equal to the thickness of the cylinder. The width of the exhaust-port is found by adding the width of the steam-port to the half-travel of the valve, and subtracting the width of the bridge.

The length of the valve is equal to the width of the exhaust-port, increased by the width of the bridges and the two faces.

V. *To find the lap and travel of the valve, and angular advance of the eccentric, for given points of admission, cut-off, and release.*—Draw a horizontal line, $A B$, and lay off on it, any distance, $A C$, to represent the length of stroke of the engine. Make $E S$ equal to the length of the connecting-rod between centres, and $S D$ equal to one half of $A C$. With D as a centre, and $D S$ as a radius, describe a circle, which represents the path described by the centre of the crank-pin. The arrow shows the assumed direction of the motion of the engine. Assume some point, E, at which it is desired to have the steam-valve begin to open when the piston has still to complete the portion $E A$ of its return stroke. Assume also a point, F, at which the steam is to be cut off when the piston has advanced a distance, $E F$, and a point, G, at which the exhaust-valve is to be begin to open when the piston has completed the portion $E G$ of the stroke. Then, with each of these points, E, F, G, as a centre, and with the length of connecting-rod, $A S$, as a radius, describe an arc of a circle cutting the path of the crank-pin in the three points e, f, g. Join each of these points with the point D, thus determining the position, $D e$, of the crank at the instant of admission, its position, $D f$, at the instant of cut-off, and its position, $D g$, at the instant of release. Bisect the angle $e D f$ by a straight line, $D L$; make $D L$ of any convenient length, and upon it, as a diameter, describe a circle, and note the point, M, in which it cuts $D e$. From D as a centre, and with $D M$ as a radius, describe an arc of a circle, $M R N$. Measure the lengths of the lines $D M$ and $D L$, and divide the former by the latter. Subtract the quotient from 1, and divide the width of steam-port by the difference. This gives the travel of the valve. Multiply the travel of the valve by half the quotient obtained above, and the product will be the steam-lap. The angle $L D I$ is the angular advance of the eccentric. Divide the length of $S R$ by the length of $L D$, and multiply half the quotient by the travel of the valve : the product is the steam-lead.

Next produce the line $L D$, making $D O$ equal to $D L$, and upon $D O$, as a diameter, describe a circle, noting the point, P, in which it cuts $D g$, the position of the crank at the instant of release. Divide the length of $D P$ by the length of $D O$, and multiply half the quotient by the travel of the valve, which gives the exhaust-lap. With D as a centre, and $D P$ as a radius, describe an arc, $P T Q$, noting the points, T and Q, in which it cuts $A B$, and the circle whose diameter is $D O$. Divide the length of $T W$ by the length of $D O$, and multiply half the quotient by the travel of the valve : the product is the exhaust-lead. Through the point Q draw the line $D h$: this is the position of the crank at the instant the exhaust-valve closes and cushion commences. With h as a centre, and $A S$, the length of the connecting-rod, as a radius, describe an arc cutting the line $A B$ in the point H; then $C H$ is the portion of the return stroke completed when the exhaust-valve closes.

On account of the angularity of the connecting-rod, the points of cut-off and exhaust closure will vary somewhat on the return stroke. They can be equalized by a slight change in the angular advance and length of the eccentric rod.

VI. *Example.*—The following example will serve to illustrate the application of the preceding principles :

A valve is to be designed for an engine having a cylinder 20 inches in diameter, and a stroke of $2\frac{1}{2}$ feet, making 80 revolutions a minute. The length of the connecting-rod is $6\frac{1}{4}$ feet. The valve is to admit steam when the piston has made 0.997 of the stroke, is to close the steam-port at two thirds of the stroke, and open the exhaust when $\frac{95}{100}$ of the stroke has been completed.

The area of the piston, in square inches, is 0.7854 times 400 (the square of 20), or 314.16, and the piston-speed is 400 feet per minute ; hence the proper port area is 0.07 times 314.16, or about 22 square inches. Assuming the length of the port to be equal to the diameter of the cylinder, 20 inches, its width will be $\frac{1}{20}$ of 22, or 1.1 inches. The width of the bridge must be equal to the thickness of the cylinder, or $1\frac{1}{8}$ inches. In the figure, make $E S$ equal to $2\frac{1}{4}$ times $A C$, and $E A$, 0.997 of $A C$, $A F$, $\frac{2}{3}$ of $A C$, and $A G$, 0.95 of $A C$. Constructing the positions of the crank corresponding to these points of stroke, and making the other constructions as explained above, suppose that the data obtained from the figure by measurement are as follows :

Angle $R D M$, 5° ; angle $M D N$, $108\frac{3}{4}$° ; angle $R D P$, $151\frac{1}{2}$° ; angle $T D Q$, $137\frac{1}{4}$° ; angle $L D I$, *angular advance of eccentric*, $40\frac{5}{8}$° ; $D M$, 0.65 of an inch ; $D L$ and $D O$, each 1.5 inches; $D P$, 0.19 of an inch ; $S R$, 0.1 of an inch ; $T W$, 0.375 of an inch ; $C H$, *part of return stroke completed when exhaust closes*, 0.85. Dividing $D M$ (0.65) by $D L$ (1.5), the quotient is 0.43+ ; subtracting 0.43 from 1, the remainder is 0.57 ; dividing twice the width of steam-port, 2.2, by 0.57, the quotient, *the travel of the valve*, is 3.86 inches. Multiplying $\frac{1}{2}$ of 0.43 by 3.86, the product, *the steam-lap*, is 0.83 of an inch. Dividing $D P$ (0.19) by $D O$ (1.5), the quotient is 0.127 ; multiplying $\frac{1}{2}$ of 0.127 by 3.86, the product, *the exhaust-lap*, is 0.25 of an inch. Dividing $S R$ (0.1) by 1.5, the quotient is 0.067 ; multiplying $\frac{1}{2}$ of 0.067 by 3.86, the product, *the steam-lead*, is 0.13 of an inch. Dividing $T W$ (0.375) by 1.5, the quotient is 0.25 ; multiplying $\frac{1}{2}$ of 0.25 by 3.86, the product, *the exhaust-lead*, is 0.48 of an inch. Adding the ha'f-travel of the valve (1.93) to the width of the steam-port, 1.1, the sum is 3.03 ; subtracting the width of the bridge (1.125), the remainder, *the width of the exhaust-port*, is 1.91 in. The sum of the steam-lap (0 83), the exhaust-lap (0.25), and the width of the steam-port (1.1), *the length of the valve-face*, is 2.18 inches. The sum of twice the length of face (4.36), twice the width of bridge (2.25), and the width of the exhaust-port (1.91), *the length of the valve*, is 8.52 inches. In this example no attempt has been made to secure great accuracy, and, as the measurements were made from a small sketch, there may be considerable errors. In practice, a scale reading to hundredths of an inch should be used, and the figure should be constructed of full size, if possible. This can generally be done by laying down the positions of the crank in a small sketch, and then transferring them to a large drawing ; or the positions may either be calculated or taken from a crank-table, if one is available. By making the drawing full size, and having marks on the shaft as described above, a template can be constructed from the drawing, for transferring the position of the eccentric to the shaft. It may be remarked, that however carefully the valve is proportioned and the adjustment of the eccentric effected, the engineer who desires to be certain that the valve-motion of his engine is properly arranged, will make the final test and adjustments with the aid of the steam-engine indicator.   B.

TESTING SMALL ENGINES.—The apparatus needed is quite simple, and can be readily constructed by the young mechanic. The following embrace the principal points that are generally of interest in regard to engines and boilers : Diameter of cylinder, length of stroke, diameters of piston-rod, connecting-rod, crankpin, valve-stem, fly-wheel, and shaft ; lengths of connecting-rod and crank-pin, weights of whole engine and of fly-wheel, size of ports, stroke of valve, point at which steam is cut off, number of revolutions per minute, clearance at each end of cylinder, pressure of steam in boiler, dimensions and weight of boiler, diameters of steam-pipe and safety-valve, number of pounds of water evaporated, fuel burned per hour, and power of the engine. Many of these data are obtained at once, by direct measurement or weight. The diameter of the cylinder should be measured when it is at the temperature at which it is ordinarily maintained while running. The point of cut-off can generally be ascertained by removing the cover of the valve-chest, and observing the point at which the steam-valve closes when the engine is moved by hand. This should be done when the parts are heated. The clearance at each end of the cylinder includes not only the space between the piston and cylinder-head at the end of the stroke, but also the volume of the ports. A simple and accurate manner of measuring the clearance is to fill the cylinder with water, when the piston is at one end of the stroke, and then measure the water carefully in a cylindrical or rectangular vessel. The difference between the volume of the water and the volume of piston displacement (area of piston multiplied by length of stroke) will be the clearance. In measuring the piston displacement at the front end of the cylinder, the volume of the piston-rod (area of section of rod multiplied by length of stroke) must, of course, be deducted.

The number of revolutions of the engine per minute can be determined approximately by observation ; but errors are apt to result, especially in the case of small engines moving at a high rate of speed. Small shaft-counters can be obtained at a very reasonable price, and measurements made with them are far more likely to be accurate.

Many small boilers are not provided with steam-gauges, so that the pressure of the steam can not be observed directly ; but all such boilers have, or should have, safety-valves, and the pressure of the steam can be determined from them. Secure the valve-stem of the safety-valve to the lever with wire or string, and attach a loop to the lever, into which pass the hook of an accurate spring-balance, arranging the loop so that it is directly over the centre of the valve-stem. Then take hold of the upper part of the spring-balance, and lift the valve slightly, noting the reading of the balance. Measure the lower diameter of the safety-valve, and find its area ; divide the reading of the spring-balance by the area of the valve, and the result will be the pressure, in pounds per square inch, at which the steam will raise the safety-valve. Suppose, for instance, that the diameter of the safety-valve is 1 inch ; its area will be about $\frac{7854}{10000}$ of an inch. Now, if the tension of the spring-balance in raising the valve is 120 lbs., the pressure at which the valve will rise is the quotient arising from dividing 120 by $\frac{7854}{10000}$, or 153 lbs. per square inch. It will be easy to make a table for any particular case, giving the pressure corresponding to each pound or fraction of a pound of tension in the balance ; and by calculating in advance the reading of the balance for any given pressure, the weight can be adjusted on the lever until that tension is obtained, and the valve can thus be graduated to lift at any required pressure. Having determined the pressure at which the safety-valve will rise when the boiler is cold, raise the valve by means of the balance, from time to time, when the engine is working, and observe the tension. Find the pressure corresponding to this tension, and subtract it from the pressure at which the valve will be raised by the steam. The difference is the pressure in the boiler at the time. For example, suppose that in the last case the tension of the balance, on raising the valve when the engine was working, was 50 lbs. The pressure corresponding to this will be 50 divided by $\frac{7854}{10000}$, or about 64 lbs., so that the pressure in the boiler at the time would be the difference between 153 and 64, or 89 lbs. per square inch. By preparing a table showing the pressure in the boiler due to each pound of tension in the spring-balance, the pressure at any time can be read off as soon as the indication of the balance is observed.

The amount of water evaporated per hour and the fuel burned can, of course, be readily determined by measurement, drawing the water from a tank of known dimensions, and observing its state at the commencement and close of a trial, being careful to leave the water in the boiler at the same height at which it was at the commencement, and maintaining this height as constant as possible during the experiment. In measuring the fuel consumed, it is best to draw out the fire at the commencement of the trial, rekindling it as soon as possible, and charging all the fuel used from that time, hauling and quenching the fire immediately at the close of the trial, and weighing back all fuel that is unconsumed. In the case of small boilers heated by lamps, a measurement of the oil used between the beginning and end of the trial will generally be sufficient ; and if gas is employed as fuel, it will be necessary to attach a meter to the pipe, to determine the quantity consumed in any given time.

To ascertain the power of the engine, the most convenient method is, generally, to attach a friction-brake, shown in the accompanying engraving, to the band-wheel. Hollow out two pieces of wood, B and C, so that they will fit the circumference of the band-wheel, A, and attach light plates of metal, D and E, to the sides, so that the pieces of wood can not slip off when secured in position. Provide two belts, F, G, countersinking the heads, H

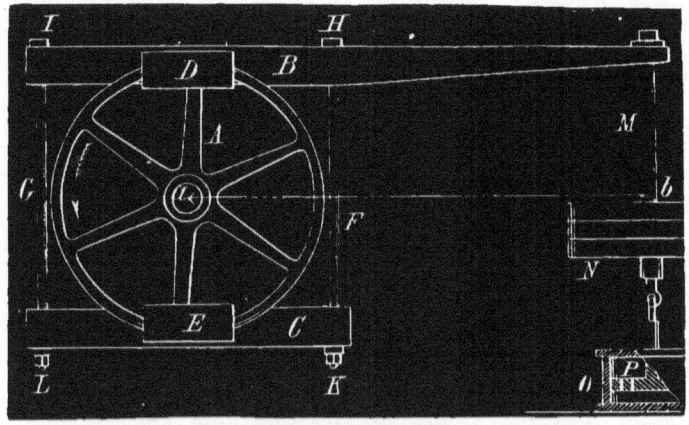

THE FRICTION-BRAKE.

and I, into the upper piece of wood, so that they can not turn, and put nuts and washers, K and L, on the other ends, so that the two pieces of wood can be clamped on the band-wheel as tightly as is necessary. Make the upper piece of wood somewhat longer than the other, and pass a rod, M, through the end. On this rod weights, N, are to be placed, and the lower end of the rod is hooked to the piston-rod of a small cylinder, O. The piston, P, fits loosely in this cylinder, which is filled with oil or water; and the piston has small holes in it, so that it can move up and down without much resistance, if moved slowly, but offers considerable resistance to sudden motion. The action of the apparatus will doubtless be apparent to our readers. By tightening the nuts on the bolts, F, G, there will be considerable friction between the band-wheel and the pieces of wood. The rod M must then be loaded with sufficient weight, so that the engine can just move at its regular rate of speed, and keep the upper piece of wood in a horizontal position. The friction on the band-wheel will cause it to become heated, unless some arrangements are made for cooling, either by keeping a stream of water running upon it, or immersing the lower part in a trough in which the water is constantly changed. The small cylinder, O, and piston, P, serve to counteract the effect of sudden shocks, which would otherwise throw the arm of the piece B from a horizontal position. Now it will be plain that, as the band-wheel revolves (constantly maintaining the arm, with the weight attached, in a horizontal position), the effect is the same as if it were lifting this weight by means of a rope running over a windlass, and the distance through which it would lift the weight in a given time is the same as the weight would move if the whole apparatus were free to revolve. If, for example, the wheel makes 300 revolutions in a minute, the distance from the centre of the wheel to the centre of the weight is 1 foot, and the weight is 10 lbs.; this weight, if free to revolve, would move in each revolution through the circumference of a circle whose radius is 1 foot, and in a minute would move 300 times as far, or about 1885 feet. The work of the engine in a minute, then, will be that required to lift 10 lbs. through a height of 1885 feet, or 18,850 foot lbs.; and as one horse-power is the work represented by 33,000 foot lbs. per minute, the engine would be developing a little more than half a horse-power.

In making experiments with the friction-brake, the apparatus should be placed loosely on the band-wheel; and before the weights are attached, a spring-balance should be secured to the arm, at the centre of the hole for the rod M, and the reading noted when the arm is in a horizontal position. This reading must be added to the weights that are afterwards attached. The horizontal distance from the centre of the wheel to the centre of the rod M, should be carefully measured. Then start the engine, with the throttle-valve wide open, and screw up the nuts K L gradually, adding weights at N. It will then only be necessary, when sufficient weights are added, to keep the wheel cool, and occasionally adjust the nuts K L, should the brake bind or become too loose from any cause. Should it be difficult or inconvenient to maintain the arm in a horizontal position, note carefully the position it assumes during the test; and for the radius to be used in the calculation, measure the distance $a\,b$ from the centre of the wheel to the centre of the rod M, in a direction perpendicular to the direction of the rod.

Instead of the weights, N, and cylinder, O, a spring-balance may be attached to the end of the rod M, and secured to some fixed support, its readings during the trial being used in place of the attached weights. In this case, also, the weight of the apparatus must be first determined, and added to the readings of the spring-balance. The plan represented in the engraving is, however, the best.

The above are, in detail, the methods to be pursued in preparing a report of the performance of small engines and boilers. Although they are far from fulfilling all the requirements of a scientific test, they will give very accurate results if carefully conducted.   B.

TURBINE WHEELS, Effective power of.—It is important, in selecting a good wheel, to be assured that it will furnish ample power. After ascertaining a reliable maker, in order to determine the exact size of the wheel it is necessary that at least one third should be allowed for variations in water levels, and for the loss consequent to the wear of wheels and gates; and, in addition, figures should be made, based on but a little more than a half gate of water to the wheel. The best wheels afford almost all of their power at a five-eighths gate or under, and a difference between a half and full gate is not more than should be the margin necessary to regulate speed. In use it will be found that opening gates seven-eighths or fully simply amounts to a large consumption of water, generally without producing five or ten per cent additional power. Some good wheels give less power when at full than at part gates. The rule should be to buy a wheel amply sufficient at not much above half gate, allowance being made for over-estimate of power. We think the experience of all who have placed wheels with a less liberal allowance will bear out and confirm this rule. Allowing one fourth for the friction of the shafting of a cotton or woolen mill, without adding one third more for a reserve when in actual use, will scarcely fail to cause a manufacturer to wish that he had bought a larger wheel. Actual tests, accurately conducted, of 31 styles of turbines show the comparative range of effective force, under the best possible advantages, to be as follows : At quarter gate, from 13 to 50 per cent ; half gate, from 11 to 71 per cent ; three-quarter gate, from 31 to 82 per cent, and at full gate, from 52 to 84 per cent, the best wheels giving out about all of their power at from five-eighths to three-quarters openings ; while the lower classes give but little power unless flooded with water, and even then fall far short of the amount claimed for them. Another reason why large wheels should be used is that, almost universally, high and low points of the head and tail waters so reduce the force of wheels as to cause partial stoppages of machinery, unless there is surplus power when the water is at the ordinary stage.

THE STEAM-BOILER AND ITS ATTACHMENTS.

BOILER, Cleaning the.—The flues or tubes of a boiler should be cleaned about once a week, with a brush or scraper. In case incrustation has formed in them, they can be cleaned by a jet of steam from a rubber hose. A boiler should be blown down and cleaned, under ordinary circumstances, about once a month. The fire should first be hauled ; and then, if possible, it is best to let the boiler stand until the water becomes tolerably cool, say for 12 hours, after which the water may be allowed to run out. Then remove the man and handhole plates, enter the boiler, and clean it with scrapers and brushes in every part that can be reached. It should then be washed out with cold water from a hose, and this washing with a hose is the only means of cleaning those parts of a boiler that can not be reached by hand. There are many boilers into which a man can not enter, and of course these can only be washed out. When the fire is hauled, all leaks in the boiler should be repaired. Leaky parts exposed to the fire must have hard patches riveted on ; in other places soft patches secured by bolts can be used, each patch having a lip around it, and the joint being made with a putty composed of red and white lead. Leaky rivets or seams can sometimes be made tight by calking. Small leaks around the ends of tubes can often be stopped in the same way, but as a general thing a leaky tube must either be replaced or plugged. To plug a tube, drive a white-pine plug tightly into each end, and cut it off even with the tube-heads ; then pass a bolt through the tube, with cup washers on each end, and screw it up tightly, putting putty under the washers.   B.

BOILERS, CYLINDRICAL.—*To find the necessary thickness in inches for the shell.*—Multiply the pressure of steam in pounds per square inch by the diameter of the boiler in inches, and multiply this product by 0.0002 for a copper boiler with single-riveted shell ; by 0.0001563 for a copper boiler with double-riveted shell ;

by 0.0001316 for a wrought-iron boiler with single-riveted shell ; by 0.0001111 for a wrought-iron boiler with double-riveted shell ; by 0.0001 for a steel boiler with single-riveted shell ; and by 0.00008333 for a steel boiler with double-riveted shell.

In illustration of the rule, suppose that it is required to find the necessary thickness for the shell of a copper boiler 60 inches in diameter, double riveted, for a pressure of 40 lbs. per square inch.

First take the product of 40 and 60, which is 2400, and multiply this by 0.0001563, which gives 0.375, or $\frac{3}{8}$ of an inch, as the necessary thickness.

To find the safe pressure in pounds per square inch.—Divide the thickness of the plate in inches by the diameter of the boiler in inches, and multiply the quotient by 5000 for a copper boiler with single-riveted shells ; by 6400 for a copper boiler with double-riveted shell ; by 7600 for a wrought-iron boiler with single-riveted shell ; by 9000 for a wrought-iron boiler with double-riveted shell ; by 10,000 for a steel boiler with single-riveted shell ; and by 12,000 for a steel boiler with double-riveted shell.

Thus, to find the safe pressure for a boiler 32 inches in diameter, the shell being made of wrought-iron plates $\frac{1}{4}$ of an inch thick, single-riveted : First divide $\frac{1}{4}$ by 32, which gives $\frac{1}{128}$, and multiply this by 7600, the product, $59\frac{3}{8}$ lbs. per square inch, being the pressure required.

Thickness, in inches, of flat heads (not stayed).—Multiply the square root of the pressure in lbs. per square inch by the radius of the shell in inches, and by 0.013333 for a head of copper ; by 0.010541 for a head of wrought-iron ; and by 0.0081649 for a head of steel.

A steel boiler has a diameter of 24 inches, and the pressure of the steam is 60 lbs. per square inch : The thickness of the head is the product of 7.746 (the square root of 60), 12, and 0.0081649, which is equal to 0.7766, or about $\frac{25}{32}$ of an inch.

Safe pressure, in pounds per square inch, for flat heads (not stayed).—Divide the square of the thickness of the plate in inches by the square of the radius of the shell in inches, and multiply the quotient by 5625 for a head of copper ; by 9000 for a head of wrought-iron ; and by 15,000 for a head of steel.

Suppose the heads of a boiler are of steel, $\frac{1}{2}$ inch in thickness, and that the diameter of the boiler is 24 inches : .25 (the square of $\frac{1}{2}$), divided by 144 (the square of the radius), is .00174, and the product of .00174 and 15,000, 26 lbs., is the pressure required. B.

BOILERS, Heating surface of.—*Note :* In the following rules all dimensions are to be taken in feet.

(*a*) *Cylindrical boilers.*—Take the product of (1) the diameter of the boiler, (2) the length of the boiler, and (3) 1.5708.

Suppose a given boiler has a diameter of 36 inches, and a length of 20 feet, its heating surface is the product of 3, 20, and 1.5708, or about $94\frac{1}{4}$ square feet.

(*b*) *Cylindrical flue boilers.*—Take the product of the diameter of the boiler, the length, and 1.5708, and add it to the product of (1) interior diameter of flue, (2) length of flue, (3) number of flues, and (4) 3.1416.

Suppose that a flue boiler is 4 feet in diameter, 22 feet long, and has two flues, each with an interior diameter of 15 inches. Then the heating surface is equal to the product of 4, 22, and 1.5708, or

Outside diameter in inches.	Outside diameter in feet.	Thickness in inches.	Internal diameter in inches.	Internal diameter in feet.	Outside surface in square feet, per foot of length.	Outside surface in square inches, per foot of length.	Internal surface in square feet, per foot of length.	Internal surface in square inches, per foot of length.
1.25	0.1042	0.072	1.106	0.0922	0.3272	47.1	0.2896	41.7
1.5	0.125	0.083	1.334	0.111	0.3927	56.5	0.3492	50.3
1.75	0.1458	0.095	1.56	0.13	0.4582	66.1	0.4084	58.8
2.	0.1667	0.095	1.81	0.1508	0.5236	75.4	0.4739	68.2
2.25	0.1875	0.095	2.06	0.1717	0.5891	84.8	0.5393	77.7
2.5	0.2083	0.109	2.282	0.1902	0.6545	94.2	0.5974	86.
2.75	0.2292	0.109	2.532	0.211	0.72	103.7	0.6629	95.5
3.	0.25	0.109	2.782	0.2318	0.7854	113.1	0.7283	104.9
3.25	0.2708	0.12	3.01	0.2508	0.8509	122.5	0.788	113.5
3.5	0.2917	0.12	3.26	0.2717	0.9163	132.	0.8535	122.9
3.75	0.3125	0.12	3.51	0.2925	0.9818	141.4	0.9189	132.3
4.	0.3333	0.134	3.732	0.311	1.0472	150.8	0.977	140.7
4.5	0.375	0.134	4.232	0.3527	1.1781	169.7	1.1079	159.5
5.	0.4167	0.148	4.704	0.392	1.309	188.5	1.2315	177.3
6.	0.5	0.165	5.67	0.4725	1.5708	226.1	1.4844	213.8
7.	0.5833	0.165	6.67	0.5558	1.8326	263.9	1.7462	251.5
8.	0.6667	0.165	7.67	0.6392	2.0944	301.8	2.008	289.2
9.	0.75	0.18	8.64	0.72	2.3562	339.3	2.262	325.7
10.	0.8333	0.203	9.594	0.7995	2.618	377.	2.5117	361.7

nearly 138¼, increased by the product of 1.25, 22, 2, and 3.1416, or about 172¾, making the total heating surface 311 square feet.

(c) *Cylindrical tubular boilers.*—Find the product of the diameter of the boiler, the length, and 1.5708, and add this to the product of the length of the boiler, the number of tubes, and the heating surface of a tube per foot of length.

The preceding table gives the heating surface per foot of length for the standard sizes of tubes.

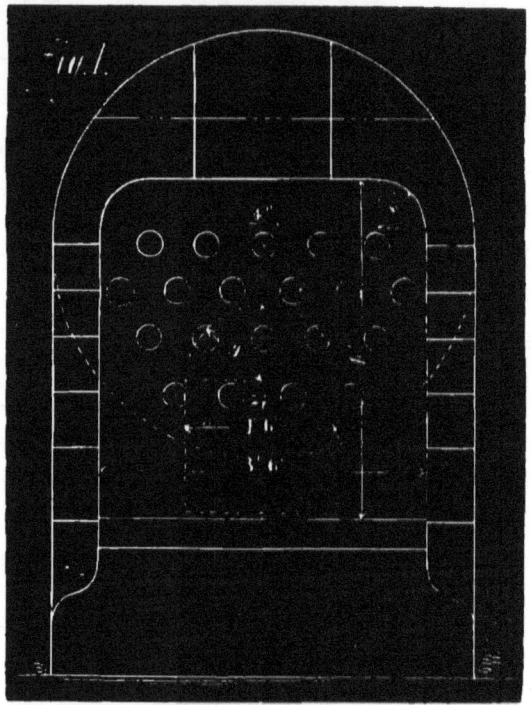

LOCOMOTIVE BOILER.

Example.—A cylindrical tubular boiler has a diameter of 42 inches, is 16 feet long, and contains 40 tubes, each 4 inches outside diameter.

The product of 3½, 16, and 1.5708 is nearly 88.

The product of 16, 40, and 0.977 (the internal surface of the tube, per running foot) is about 615, so that the whole heating surface is 703 square feet.

(d) *Locomotive boilers.*—I. Add together the following quantities: (1) The product of the length of the line bounding the cross-section of the furnace, and the length of the furnace. (2) Twice the area of the cross-section of the furnace. (3) The product of the length of the tubes, the number of tubes, and the heating surface of a tube per foot of length.

II. Subtract from this sum the sum of the following quantities: (4) The area of the furnace-door. (5) The product of the number of tubes, the square of the internal diameter of a tube, and 0.7854.

As an example of the use of this rule, suppose it is required to determine the heating surface of a boiler having the dimensions noted in the engravings, Fig. 1 being a cross-section of the

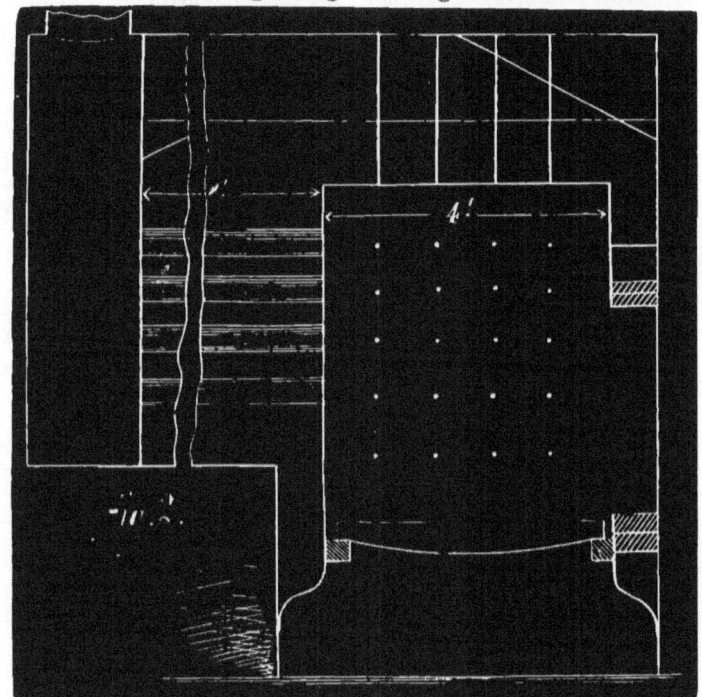

LOCOMOTIVE BOILER.

boiler at the furnace, showing also the furnace door in dotted outline, and Fig. 2 a longitudinal section. (1) The length of the line bounding the cross-section of the furnace is the sum of twice 3.5, 1.5708, and 2.5, or 11.07, and the product of 11.07 and 4 is 44.28. (2) The area of the cross-section of the furnace is the sum of 3.5 squared, $2\frac{1}{2}$ times $\frac{1}{2}$, and 0.7854 divided by 2, or about 13.89. Twice 13.89 is 27.78. (3) The product of 8, 20, and 0.977 is 157.32. The sum of 44.28, 27.78, and 157.32 is 229.38. (4) The area of the furnace-door is the sum of 1.5 times 1.25, and 0.3927 times 1.5 squared, or about 2.76. (5) The product of 20, 0.311 squared, and 0.7854, is about 1.52.

The sum of 2.76 and 1.52 is 4.28. The difference between 229.38 and 4.28 is about 225 square feet, the heating surface required.

(e) *Vertical boilers.*—I. Take the sum of the following quantities : (1) The product of the diameter of furnace, height of same, and 3 1416. (2) The product of the diameter of the furnace squared and 0.7854. (3) The product of the number of tubes, length of same, and heating surface per foot of length.

II. Subtract from this sum the product of the number of tubes, the internal diameter of a tube squared, and 0.7854.

Example.—Required, the heating surface of a vertical boiler with the following dimensions : Diameter of furnace, 24 inches ; height of furnace, 18 inches ; 40 tubes, each 2 inches outside diameter, 6 feet long.   (1) The product of 2, 1.5, and 3.1416 is 9.42.   (2) The product of 4 and 0.7854 is 3.14.   (3) The product of 40, 6, and 0.4739 is 113.74.   The sum of 9.42, 3.14, and 113.74 is 126.3.   The product of 40, 0.02274 (the square of 0.1508), and 0.7854 is about 0.72.   The heating surface is the difference between 126.3 and 0.72, which is about 125.6 square feet.   B.

BOILERS, HORIZONTAL, Setting.—The best way is to have the fire-box at least as wide as the boiler, and have as much heating surface as possible ; but below the water-line all passages should be made large, so as to allow a free passage to the heated gases, and where they leave the boiler, a damper should be provided. The bridge-wall should be high enough to prevent coal from being thrown over, and the grates low enough to allow ample room for combustion.   Nothing can be gained by putting the fire near the boiler or contracting any of the passages ; it is better to let the heat diffuse itself fully throughout the entire heating surface.

BOILERS, Priming in.—If your boiler primes, either "swap" it off for another or superheat your steam moderately ; but beware of anti-priming doctors and their remedies.

BOILERS, Rules for firing under.—(1) Begin to charge the furnace at the bridge end, and keep firing to within a few inches of the dead-plate.   (2) Never allow the fire to be so low before a fresh charge is thrown in, so that there shall be at least 4 or 5 inches of clear, incandescent fuel on the bars, equally spread over the whole.   (3) Keep the bars constantly and equally covered, particularly at the sides and bridge end, where the fuel burns away most rapidly.   (4) If the fuel burns unequally or in holes, it must be leveled and the vacant spaces filled.   (5) The large coals must be broken in pieces not larger than a man's fist.   (6) Where the ash-pit is shallow, it must frequently be cleaned out ; a body of hot cinders will overheat and burn the bars.

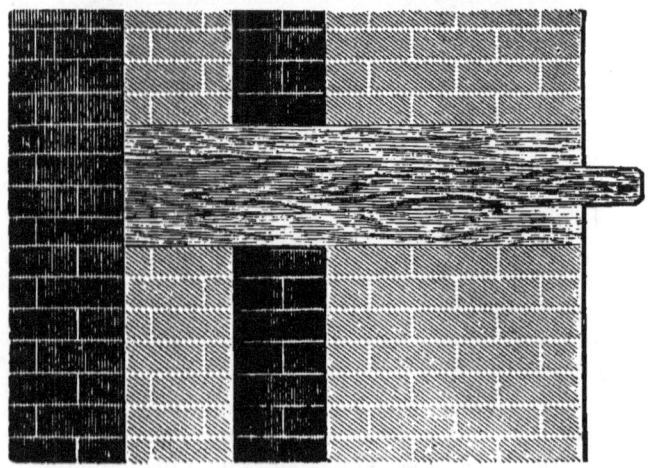

STRAIGHTENING TALL CHIMNEYS.

CHIMNEYS, To straighten tall.—Have a number of oak wedges made of sufficient length to pass through the entire thickness of the chimney and project sufficiently on the outside. Place them in sets of three each, one over the other, as shown in the engraving, having the surfaces in contact straight and smooth, and black-leaded to diminish friction. Commence on the opposite side to that in which the chimney leans; cut through to the inside, insert one set of wedges, and wedge above and under them until they take a bearing. Repeat the process around the chimney, except on the lowest side, leaving spaces of a foot or more between each set of wedges. Then, by driving the centre wedge in each set inwards, as much of the chimney as rests on them is gradually lowered just at the places and to the amount required to bring it to an exact perpendicular. When that is done, brick up the intervening spaces, loosen and withdraw the wedges, and brick up in their places. This requires careful and skilful work.

CHIMNEYS, Proportioning.—The general rule is to make the cross-section of the chimney, which may be either round or square, from $\frac{1}{8}$ to $\frac{1}{10}$ of the grate-surface, and the height from 50 to 70 feet.

To determine the amount of coal which will be burned per square foot of grate per hour, with good proportions, by Professor Thurston's rule.—Subtract one from twice the square root of the height. *Example:* What will be the amount of coal burned per square foot of grate surface per hour, the chimney being 49 feet high, and suitably proportioned? The square root of 49 is 7; twice 7 is 14; 14 less 1 is 13, which is the amount of coal required in pounds.

To determine the height required to give a certain rate of combustion.—Add 1 to the weight to be burned per square foot per hour; divide by 2 and square the quotient. *Example,* same as above, worked backwards, thus: What height of chimney is required to burn 13 pounds of coal per square foot of grate surface per hour? 13 and 1 are 14; 14 divided by 2 is 7; the square of 7 is 49, which is the height of the chimney in feet.

COAL, Effect of damp air on.—It has been found by recent experiments on this subject, that the loss in weight, due to a slow oxidation and to the disengagement of gases which form the richest part of bituminous coal, may equal one third of the original weight. The heating power in such coal was lowered to 47 per cent of its former capacity. The same coal exposed to the air, but in a closed receptacle, did not lose more than 25 per cent of gas and 10 per cent of heating power. Bituminous coals alter most rapidly. This shows the disadvantage of damp cellars, and of leaving coal uncovered for long periods and subject to bad weather.

COMBUSTION AND FUEL.—The principal constituents of coal are carbon, hydrogen, water, a little sulphur, ashes and clinker, the latter two substances consisting generally of silica, alumina, iron, lime, magnesia, and oxide of manganese. The principal combustible constituent of anthracite is fixed, uncombined carbon. The free-burning or semi-bituminous coals contain a considerable amount of hydrocarbon or volatile combustible matter, and bituminous caking coals have a larger percentage of volatile combustible.

If a mass of coal is brought to a sufficiently high temperature (probably something above 1000° Fahr.), the combustible materials enter into chemical combination, and as much heat is given out as would be required to decompose the resulting products into their elements. When coal is burned the water is first expelled; then the sulphur, if any is contained, is consumed, forming sulphurous oxide; after this the hydrogen in the volatile combustible matter unites with oxygen, forming water; and the carbon set free unites with oxygen, forming carbonic dioxide, if the temperature is sufficiently high and enough oxygen is present, or, under less favorable circumstances, either forming carbonic oxide or passing off unconsumed, as soot. The combustion of the fixed carbon next begins, the product of the combustion being carbonic dioxide or carbonic oxide, so that finally nothing is left except the ashes and clinker.

It may be well to trace the effect of these various combinations: The water contained in the coal is expelled in the form of steam, so that it carries off some heat, and is a positive disadvantage. The complete combustion of a pound of sulphur produces about 4000 units of heat, but the amount of sulphur in coal is usually so slight that its heating qualities scarcely deserve to be regarded. The action of the sulphur on the material of the boiler is, however, a very serious matter. It has not yet been determined by experiment what per cent of sulphur is sufficient to render a coal unfit for use in a furnace, but it is well known that many of the Western coals produce very bad effects when employed in locomotive boilers. A pound of hydrogen combining with oxygen, forming 9 pounds of water, has a heating power of 62,032 units. It seems doubtful, however, whether this amount of heat is available from the combustion of hydrogen in a boiler. The experiments by which this value was determined were made upon hydrogen in the gaseous state, and the steam resulting from the combustion was condensed. Now the hydrogen in coal is ordinarily combined with carbon, and frequently with nitrogen, so that it must be separated from the combination before it can be united with oxygen, and heat is required for this separation. Again, in a boiler the products of combustion usually pass into the chimney at such a high temperature that the water, which is the result of the combustion of the hydrogen, passes off in the form of steam, and thus carries off a considerable quantity of what is commonly known as latent heat. This subject is one which has been but little considered by experimenters, and is worthy of more extended investigation. The volatile combustible matter of coal generally contains oxygen in combination, and this must be changed into the gaseous state before being united with the hydrogen, an operation that requires as much heat as results from the new combination. Hence it is certain that the amount of hydrogen contained in coal must be diminished by one eighth of the weight of the oxygen before attempting to form any estimate of its heating qualities.

The carbon of the coal, as has been already stated, will unite with oxygen, forming carbonic dioxide, and may afterwards take up more carbon, and be converted into carbonic oxide. Now the result of the complete combustion of a pound of carbon is $3\frac{2}{3}$ lbs. of carbonic dioxide, and the combustion produces 14,500 units of heat. But a pound of carbon imperfectly burned produces $2\frac{1}{3}$

| KIND OF COAL. | PERCENTAGE. | | | Weight, in pounds, per cubic foot. | Pounds of water evaporated, from and at 212°, by one pound of fuel. | Pounds of water evaporated, from and at 212°, by one cubic foot of fuel. | Pounds of water evaporated, from and at 212°, by one pound of combustible. | Per cent of waste, in clinker and ashes. | Per cent of clinker. |
	Sulphur.	Fixed carbon.	Volatile combustible matter.						
Anthracite from Pennsylvania.	0.052	88.543	3.981	53.30	9.55	509.9	10.47	8.64	1.70
Semi-bituminous from Maryland.	0.714	73.952	14.198	53.14	9.98	530.1	11.08	10.71	3.13
Semi-bituminous from Pennsylvania.	0.722	72.469	16.019	52.55	9.43	493.9	10.71	11.84	3.10
Bituminous caking coal from Virginia.	1.232	58.104	29.432	49.27	9.52	414.6	9.52	10.95	5.39
Bituminous from Nova Scotia.	0.769	61.762	25.616	50.08	8.29	415.8	9.29	10.48	4.86
Bituminous from Great Britain.	0.321	53.569	38.007	49.93	7.82	389.6	8.38	6.94	3.54
Bituminous from Cannelton, Indiana.		58.437	33.992	47.65	7.34	348.8	7.73	5.12	1.64
Dry pine wood.				21.00	4.69	98.6	4.71		

pounds of carbonic oxide, and only 4400 units of heat. In a furnace where the combustion is imperfect, the action is usually as follows : A pound of carbon is at first completely burned, forming 3⅔ pounds of carbonic dioxide, and then takes up another pound of carbon, producing 4⅔ pounds of carbonic oxide and 8800 units of heat. This carbonic oxide, however, if supplied with a sufficient amount of air, will burn and again form carbonic dioxide, so that the full effect of the combustion of the carbon will be realized. A practical application of this principle is seen in the combustion-chambers in boilers, which are designed to complete the combustion of the gases after they leave the furnace.

Having disposed of the materials of the coal which escape into the chimney of a boiler, the ashes and clinker that remain should be considered. The effect of these substances is injurious in several ways : they choke up the furnace, preventing free access of the air to the combustible materials, and instead of entering into combinations and producing heat, they require to be heated to the temperature of the furnace, and are then removed, without having produced an equivalent for the heat expended upon them.

In ordinary boiler-furnaces, the amount of air required for the combustion of 1 pound of coal is about 24 pounds, or between 280 and 300 cubic feet.

For a table of the qualities of American coal from various localities, compiled from Prof. Johnson's Report, see page 95. B.

FEED-WATER HEATERS, Gain from the use of.—A unit of heat is the amount of heat required to raise the temperature of a pound of water one degree, the water being at the temperature of maximum density, about 39.1° Fahrenheit. The table below shows the number of units of heat required to convert one pound of water, at the temperature of 32°, into steam of various pressures.

Pressure of steam in lbs. per sq. in., by gauge.	Units of heat.	Pressure of steam in lbs. per sq. in., by gauge.	Units of heat.
1	1148	10	1155
20	1161	30	1165
40	1169	50	1173
60	1176	70	1178
80	1181	90	1183
100	1185	110	1187
120	1189	130	1190
140	1192	150	1193
160	1195	170	1196
180	1198	190	1199
200	1200		

In a non-condensing engine, if the exhaust steam escapes directly into the atmosphere, it carries off most of the heat that was previously imparted to it by the coal in the furnace. This may be illustrated by an example.

Suppose the steam is admitted into the cylinder of an engine at a pressure of 90 pounds per square inch, and exhausted at a pressure of 1 pound above the atmosphere, and that the temperature of the feed-water is 70°. If the feed-water had been at 32°, it will be seen that each pound of water would have required 1183 units of heat to convert it into steam of 90 pounds pressure ; but since the temperature of the feed was 38° above 32°, 38 units less, or 1145, will be necessary. The exhaust steam carries away 1148 less 38, or 1110 units of the heat that has been imparted to the steam by the coal, so that about 96.94 per cent of the heat is thrown away. Now, it costs money to heat this water, and the steam-user who pays $6.50 a ton for coal, which converts 15,000 pounds of water into steam, might make up an account somewhat after this manner : "After evaporating 15,000 pounds of water at a cost of $0.00043+ per pound, I allowed 14,541 pounds to escape into the air without rendering me any equivalent, and only utilized 459 pounds in my engine ; so that really I pay at the rate of $0.014+ for every pound of water used."

There are many steam-users to-day who could readily make up an account somewhat like the preceding.

Now suppose that the steam-user, being convinced of the folly of paying for coal to raise steam which is blown away without doing any good, attaches a heater to his exhaust-pipe, and so raises the temperature of the feed-water to 200°, instead of 70°, as before. If this is done, and the heater is a good one, which does not increase the back pressure, each pound of water requires 130 units less for its evaporation, and each pound of exhaust-steam carries away 130 units less than before ; and the steam-user can make up his account anew, as below:

"One ton of coal now evaporates about 16,900 pounds of water, at a cost of $0.00038 per pound, and 600 pounds of this water are utilized in the engine ; so that I now pay $0.0108 for each pound of water that produces useful effect."

If he had an engine of 100 horse-power, using 30,000 pounds of steam a day, and working 300 days in a year, he would find that the difference in his coal bill, before and after the change, would be the difference between $3870 and $3420, or $450.   B.

FIRE-CLAY for boiler-furnaces.—Take common earth, well mixed with water, to which is added a small quantity of rock-salt ; let the water stand until the salt dissolves, which will take about 2 or 3 hours. It is then ready for use. Apply it as fire-clay is used, and your furnace will stand much longer.

FLUE-SHEETS, To calk leaks in.—Use a reflector (a tin plate will do) adjusted in front of the furnace-door, so as to throw light on to the flue-sheets, while calking leaks.

FLUES, WROUGHT-IRON.—*Thickness in inches of a wrought-iron flue exposed to external pressure :* I. Find (1) the product of the diameter of the flue in inches, (2) the length of the flue in feet, (3) the pressure of steam in lbs. per square inch, and (4) .000,009,091.

II. Extract the square-root of this product.

Suppose, for example, that a flue is 12 inches in diameter and 6 feet long, and that the pressure of steam is 40 lbs. per square inch : The product of 12, 6, 40, and 0.000,009,091 is about 0.02618, and the square-root of this number is 0.1618+, or about $\frac{3}{32}$ of an inch. It will be observed that the thickness of a flue increases directly as the length. Thus, other things being equal, a flue that is 12 feet long must be twice as thick as one that has a length of 6 feet. In making long flues, it is common to strengthen them at intervals by bands, thus converting them into a series of short flues, so far as the strength is concerned.

Safe pressure in lbs. per square inch for a wrought-iron flue exposed to external pressure : Multiply the square of the thickness in inches by 110,000, and divide this product by the product of (1) the diameter of the flue in inches, and (2) the length of the flue in feet.

Example.—What is the safe pressure for a flue 15 inches in diameter, 8 feet long, and ⅜ of an inch thick ?

The product of 0.140,625 (the square of the thickness) and 110,-000 is 15,468.75. Dividing this by 15 times 8, or 120, the quotient is about 128.9 lbs., the pressure required.   B.

GAUGES, mercurial steam, Keeping clean.—This can be done by putting a little glycerine or sulphuric acid on the surface of the mercury. This serves as a lubricator of both glass and metals, and prevents their adhesion.

GAUGES, Water and steam.—When a boiler is in use, the gauge-cocks should frequently be tried to see that they are not choked up, and the glass gauge should often be blown out. After ascertaining the proper place for the weight on the lever of the safety-valve, a stick should be secured to the lever with wire, so that the ball can not be moved out any further. A cord should be secured to the safety-valve lever, within easy reach of the engineer, so that the valve can be opened by hand if it sticks, and the safety-valve should be tried at least once every day, to ascertain whether or not it is in working order.

A steam-gauge should be tested at least once a year, and the engineer should frequently try its accuracy by allowing the steam to raise the safety-valve, and noting the pressure shown by the gauge. The hand of a steam-gauge sometimes sticks, and the engineer should tap the face of the gauge lightly several times a day, to assure himself that it is in working order. He may also shut off the steam from the gauge-pipe, and open the drip-cock, noting whether the hand goes back promptly to 0, and returns to the former reading when steam is again turned on.

In testing a boiler, warm water should be used, and a better test, when this is possible, is to enter the boiler and make a thorough internal examination.

In leaving a boiler for the night, the fire may either be hauled or banked. If it is to be banked, it should first be cleaned and then pushed back and covered with coal, the boiler being left with the furnace-door open and the damper closed.   B.

JOINTS, RIVETED, Proportions of.—(*a*) *Diameter of rivet in inches :* Multiply the thickness of plate in inches by 2 for plates up to ⅜ of an inch thick, by 1.5 for plates from ⅜ to ⅝ of an inch thick, by 1.25 for plates from ⅝ to ¾ of an inch thick, and by 1.125 for plates from ¾ to 1 inch thick.

(*b*) *Length of rivet under the head, in inches :* Multiply the thickness of the plate in inches by 4.5.

(*c*) *Distance between rivets, from centre to centre, in inches :* (1) *Single-riveted joints :* Multiply the thickness of plate in inches by 6 for plates up to ¼ of an inch thick, by 5 for plates from ¼ to ⅜ of an inch thick, by 4 for plates from ⅜ to ⅝ of an inch thick, and by 3 for plates from ⅝ to 1 inch thick.

(2) *Each line of rivets, double-riveted joints :* Multiply the thickness of the plate in inches by 7 for plates up to ¼ of an inch thick,

by 6 for plates from $\frac{1}{4}$ to $\frac{7}{16}$ of an inch thick, by 5 for plates from $\frac{7}{16}$ to $\frac{9}{16}$ of an inch thick, and by 4 for plates from $\frac{9}{16}$ to 1 inch thick.

(*d*) *Lap to be given to joint, in inches:* (1) *Single-riveted joints :* Multiply the thickness of plates in inches by 6 for plates up to $\frac{3}{8}$ of an inch thick, by 4.5 for plates from $\frac{3}{8}$ to $\frac{3}{4}$ of an inch thick, and by 4 for plates from $\frac{3}{4}$ to 1 inch thick.

(2) *Double-riveted joints :* Multiply the thickness of plate in inches by 10 for joints up to $\frac{3}{8}$ of an inch thick, by 7.5 for joints from $\frac{3}{8}$ to $\frac{3}{4}$ of an inch thick, and by 6.7 for joints from $\frac{3}{4}$ to 1 inch thick.

A table is appended, giving the thickness of plate in decimals of an inch, varying by sixteenths :

Halves.	Fourths.	Eighths.	Sixteenths.	Decimals.
			1.........	0.0625
		1.......	2..........	0.125
			3..........	0.1875
	1.........	2.........	4	0.25
			5	0.3125
		3.........	6..........	0.375
			7..........	0.4375
1.........	2.........	4.........	8..........	0.5
			9..........	0.5625
		5.........	10.........	0.625
			11.........	0.6875
	3........	6.........	12.........	0.75
			13.........	0.8125
		7.........	14.........	0.875
			15.........	0.9375
2.........	4.........	8.........	16...... ..	1.

The following example will serve to illustrate the rules :

What should be the proportions of a single-riveted joint for a boiler made of plates $\frac{1}{8}$ of an inch thick ?

Diameter of rivets is twice 0.125, or $\frac{1}{4}$ of an inch. Length of rivets under head is 4.5 times 0.125, or $\frac{9}{16}$ of an inch. Distance between centres of rivets is 6 times 0.125, or $\frac{3}{4}$ of an inch. Lap of joint is 6 times 0.125, or $\frac{3}{4}$ of an inch. **B.**

LEAD, Effect of steam on.—Lead in contact with steam under pressure of over 10 lbs. per square inch very soon loses its strength, and it is therefore good neither for packing joints nor for conveying steam.

PIPES, STEAM, Burst.—Steam-pipes which have cracks in them from having burst, may be repaired by heating and then soldering them.

PIPES, STEAM, Condensation in subterranean.—To prevent this, inclose the pipe in another larger pipe, and fill the space between the two with plaster-of-Paris or charcoal. The outside pipe should be water-tight.

PIPES, STEAM, Isolating material for.—Take 132 lbs. limestone, 385 lbs. coal, 275 lbs. clay, and 330 lbs. sifted coal-ashes. This is finely pulverized, and mixed with 660 lbs. water, 11 lbs. sulphuric acid at 50° B., and about 160 lbs. calves' hair or hog-bristles. The compound is applied to the pipes in coats of 0.4 inch thickness, repeated until a thickness of an inch and a half is obtained, when a light covering of oil is given.

PIPES, STEAM, To prevent cracking, from freezing.—Steam-pipes apt to fill with condensed water and burst from freezing should have small holes with plugs to them, the plugs to be taken out at night.

SAWDUST AS FUEL, To burn.—A saw-mill owner solves the problem of using saw-dust as fuel as follows. His boiler was a return tubular, 14 feet 6 inches long and 54 inches in diameter, with 64 three-inch tubes, and brick firebox 48 x 56 x 27 inches high; bridge-wall was 7 inches at centre, rounded to the sides of boiler. He states: "I lowered the bridge-wall 13 inches (keeping the same circle as before), lowered the paving in rear of firebox to a level with the grate-bars, and obtained a barrel of furnace-slag from 3 to 7 or 8 inches in size and 1 or 1½ inches thick, which I placed on the grate-bars, about half covering them. I fired with wood; and when the slag got heated, I threw in the sawdust, which burned very well but smoked fear-fully (clouds would arise from the smoke-stack). I then intro-duced a 2-inch pipe, with about fifty ½-inch holes, directly behind the bridge-wall, leaving both ends of pipe open; after which, I never had a particle of trouble either in keeping up steam or in burning up the smoke. Not even in firing up did I ever see any smoke come out of the stack, which was 30 feet high and 32 inches square, enlarged near top and to the top to 36 inches in-side measurement. I forgot to state that I covered the top of boiler with sheet-iron, then laid brick on it, covering the inter-stices with sand. The sheet-iron was to prevent the sand from wedging off the wall when the boiler expanded." A system of alternate firing, and with grates so arranged as to permit some charred fuel to fall through and burn in the ash-pit, gives the best results.

SCALE IN STEAM-BOILERS, Prevention of.—(1) Use as pure water as your locality affords. (2) Clean and scrape your boiler as often as you possibly can. (3) Blow off without excess. (4) In case of salt or brackish waters, never use steam of over 90 lbs. pressure to the sq. in. (5) In case of sulphate of lime waters, never use steam of over 70 lbs. pressure. (6) In case of water holding carbonate of lime in solution, pass it through a feed-water heater made hot by exhaust steam or waste heat. (7) In case of muddy waters use large feed-water cisterns or reservoirs, on the bottom of which the suspended earthy matters will soon form a soft deposit, when the surface water can be drawn off for use. When using hard water, save the drippings of the exhaust-pipe, and the condensation of the safety-valve blow-off, and from the cylinder, and use the water thus obtained to fill the boiler after blowing off. The result will be surprising in its ef-fect in loosening scale.

SLACK, To burn as fuel.—A correspondent, who has practical-ly investigated this subject extensively, says: Slack requires the grate-bars to be very open. I have used bars with openings of 1¼ inches. The only secret in using slack with any kind of a furnace is to have the grate-bars open enough so that the fire can be kept open from the under side of the grates with the poker. Some coal, of course, will go through at first ; but coarse coal and wood can be used to start with, and what falls through the grate must be raked out and put in again. The coal will soon cake so that it will not waste. Make the furnace wider than usual, in building it, with doors in the side of the front, similar to furnaces for burning sawdust. For some varieties of coal, it will be found beneficial to wet the coal before throwing it into the furnace ; this helps to run the coal together. Then put in the coal at the side doors, and let it alone till it cakes ; then with the poker roll it into the centre of the fire. It will be in large lumps and will not waste, and there will always be a good fire in the centre. Never smother it with fresh coal. A system of introducing comminuted fuel with the air required for its combustion, by means of a fan-blower, has been introduced by a Boston firm.

SPECIFIC HEAT.—*Table showing the number of units of heat required to raise the temperature of one pound of a substance one degree Fahrenheit.*

Air	0.23740	Copper	0.09515
Alcohol (liquid)	0.61500	Corrosive sublimate	0.06889
" (vapor)	0.45340	Corundum	0.19762
Aluminum	0.21430	Diamond	0.14687
Ammonia (vapor)	0.50830	Ether (liquid)	0.50342
Anthracite coal	0.20100	" (vapor)	0.48100
Antimony	0.05077	Fusel Oil	0.56400
Aragonite	0.20850	Galena	0.05086
Arsenic	0.08140	Glass	0.19768
Benzine	0.45000	Glucinum	0.23080
Bismuth (solid)	0.03084	Gold	0.03244
" (liquid)	0.03630	Graphite	0.20083
Bituminous coal	0.20085	Hydrochloric acid	0.18450
Boron	0.25000	Hydrogen	3.40900
Brass	0.09391	Ice	0.47400
Bromine (liquid)	0.10700	Iceland spar	0.20858
" (gas)	0.05550	Indium	0.05700
Cadmium	0.05669	Iodide of mercury	0.04197
Carbonic acid	0.21630	" " potassium	0.08191
" oxide	0.24500	" " silver	0.06159
Chalk	0.21485	Iodine (solid)	0.05412
Charcoal	0.24150	" (liquid)	0 10822
Chloride of barium	0.89570	Iridium	0.03259
" " calcium	0.16420	Iron	0.11380
" " lead	0.06641	Iron pyrites	0.13001
" " magnesium	0.19460	Lead (solid)	0.03065
" " manganese	0.14250	" (liquid)	0.04020
Chloride of strontium	0.11990	Lithium	0.94080
" " zinc	0 13618	Magnesium	0.24990
Chlorine (gas)	0.12100	Manganese	0.12170
Chromium	0.12000	Marble	0.20989
Cobalt	0.10730	Mercury (liquid)	0.03332

Mercury (solid)	0.03192	Selenium	0.07446
Molybdenum	0.07218	Silica	0.19132
Nickel	0.11080	Silicon	0.17740
Niobium	0.06820	Silver	0.05701
Nitrate of sodium	0.27821	Sodium	0.29340
" " silver	0.14352	Steam	0.48050
Nitre	0.23875	Steel	0.11750
Nitric oxide	0.23150	Sulphide of carbon	0.15700
Nitrogen	0.24380	" " zinc	0.12813
Nitrous oxide	0.22380	Sulphur (native)	0.17760
Oil of turpentine (liq'd)	0.46727	" (purified)	0.20259
" " " (vapor)	0.50610	" (liquid)	0.23400
Olefiant gas	0.40400	Sulphuric acid	0.34300
Olive oil	0.31000	Tantalum	0.04840
Osmium	0.03113	Tellurium	0.04737
Oxygen	0.21750	Thallium	0.03355
Palladium	0.05928	Thorinum	0.05800
Petroleum	0.46840	Tin (solid)	0.05623
Phosphorus	0.18870	" (liquid)	0.06370
Platinum	0.03243	Tungsten	0.03342
Potassium	0.16956	Uranium	0.06190
Rhodium	0.05803	Vanadium	0.08140
Ruthenium	0.06110	Water	1.00000
Salt	0.17295	Wood spirit	0.64500
Sapphire	0.21737	Zinc	0.09555
			B.

STAYED SURFACE, Safe pressure, in lbs. per square inch, for a.—Divide the square of the thickness of the plate in inches, by the square of the distance between stays, in inches, and multiply the quotient by 16,875 for a copper plate, by 27,000 for a wrought-iron plate, and by 45,000 for a steel plate

Example.—What is the safe pressure for a plate of wrought iron, ¼ of an inch thick, secured by stays 6 inches from centre to centre?

The quotient arising from dividing 0.0625 (the square of ¼) by 36, is 0.00174. Multiplying 0.00174 by 27,000, the product is the required pressure, about 47 lbs. per square inch.　　　　B.

STAYED SURFACE, Thickness of, in inches.—Multiply the square root of the pressure, in lbs. per square inch, by the distance between centres of stays in inches, and multiply this product by 0.007698 for a copper plate, by 0.0060858 for a wrought-iron plate, by 0.0047141 for a steel plate.

For a copper fire-box, in which the stays are 10 inches apart from centre to centre, and the pressure of steam is 60 lbs.: The thickness of plate is the product of 7.746 (the square root of 60), 10, and 0.007698 ; which is equal to 0.596, or about ⅓⅜ of an inch.　　　　B.

STAY, Proper diameter for a, in inches.—Multiply the distance between stays, in inches, by the square root of the pressure, in pounds per square inch, and multiply this product by 0.0206 for a copper stay, by 0.01784 for a wrought-iron stay.

Example.—What is the proper diameter for wrought-iron

stays, 6 inches between centres, the pressure of steam being 75 pounds per square inch ?

This is the product of 6, 8.66 (the square root of 75), and 0.01784 ; which is equal to 0.92697, or about $\frac{1}{16}$ of an inch. B.

STAYS, Distance between, in inches.—Divide the thickness of the plate, in inches, by the square root of the pressure, in lbs. per square inch, and multiply the quotient by 130, if the stayed surface is copper ; by 164, if the stayed surface is wrought iron ; by 212, if the stayed surface is steel.

Suppose the fire-box of a boiler is to be made of steel plates, $\frac{3}{8}$ of an inch thick, and the pressure of steam is to be 100 lbs. per square inch.

Divide 0.375 by 10, and multiply the quotient, 0.0375, by 212 ; which gives 7.95, say 8 inches, as the proper distance between stays. B.

Note.—The rules for stayed surfaces and flat boiler-heads are adapted from methods explained by Dr. Grashof in " Die Festig-keitslehre," Berlin, 1866.

VALVE, SAFETY, A simple test for determining the accuracy of.—Secure the valve-stem of the safety-valve to the lever with wire or string, and attach a loop to the lever, into which pass the hook of an accurate spring-balance, arranging the loop so that it is directly over the centre of the valve-stem. Then take hold of the upper part of the spring-balance, and lift the valve slightly, noting the reading of the balance. Measure the lower diameter of the safety-valve, and find its area ; divide the reading of the spring-balance by the area of the valve, and the result will be the pressure, in pounds per square inch, at which the steam will raise the safety-valve. Suppose, for instance, that the diameter of the safety-valve is 1 inch ; its area will be about $\frac{7854}{10000}$ of an inch. Now, if the tension of the spring-balance in raising the valve is 120 lbs., the pressure at which the valve will rise is the quotient arising from dividing 120 by $\frac{7854}{10000}$, or 153 lbs. per square inch.

A table is appended, giving the areas of valves for the majority of cases that occur in practice :

Table of Areas of Valves of Different Diameters.

Diameter of valve in inches.	Area of valve in square inches.
$\frac{1}{2}$ or 0.5	13-64 or 0.19635
$\frac{5}{8}$ or 0.625	5-16 or 0.30680
$\frac{3}{4}$ or 0.75	7-16 or 0.44179
$\frac{7}{8}$ or 0.875	19-32 or 0.60132
1	25-32 or 0.7854
$1\frac{1}{4}$ or 1.25	1 15-64 or 1.2272
$1\frac{1}{2}$ or 1.5	1 49-64 or 1.7671
$1\frac{3}{4}$ or 1.75	2 13-32 or 2.4053
2	3 9-64 or 3.1416
$2\frac{1}{2}$ or 2.5	4 29-32 or 4.9087
3	7 1-16 or 7.0686
$3\frac{1}{2}$ or 3.5	9 5 8 or 9.6211
4	12 9-16 or 12.5664
$4\frac{1}{2}$ or 4.5	15 29-32 or 15.9043

Diameter of valve in inches.	Area of valve in square inches.
5	...............19 41-64 or 19.635
5½ or 5.5	...............23 49-64 or 23.7583
6	...............28 9-32 or 28.2744

B.

VALVE, SAFETY, Hints concerning the.—Some convenient arrangement, such as a cord or lever, should be fitted to a safety-valve, so that it can readily be opened by hand ; and the valve should be moved at least once a day, to keep it in good working order. A simple experiment to determine whether or not the valve is in truth a *safety* valve can readily be made by every steam-user. It will only be necessary to shut off the steam from the engine, or wherever else it is used, and making up a good fire in the boiler, observe whether the pressure increases materially beyond the point for which the valve is set. This experiment can be made without the slightest danger, since, if the valve will not relieve the boiler automatically, it can be opened to any desired extent by hand. Any one can readily perceive the importance of making this test, for with a good safety-valve in working order, the chances of a disastrous boiler explosion are greatly diminished. B.

VALVE, SAFETY, Proper diameter, in inches, for a.—This depends upon (1) the steam-pressure to which the valve is exposed ; (2) the lift of the valve ; (3) the quantity of steam that must be discharged in a given time, in order to prevent an increase of pressure. These quantities having been determined, it is necessary to calculate (1) the area of opening required in order to discharge the given quantity of steam ; (2) the diameter of a valve that will afford the required area of opening with the given lift.

The method of making these calculations is explained below.

A. *The area of opening, in square inches, required, in order that a safety-valve may prevent the increase of steam-pressure beyond a given point.*

(*a*) For stationary and marine boilers with natural draft : Take the product of (1) the area of the grate-surface in square feet, and (2) 2.63, and divide this product by the steam-pressure as shown by gauge, increased by 14.7.

(*b*) For stationary and marine boilers with forced draft : Take the product of (1) the area of the grate surface in square feet, and (2) 4.08, and divide this product by the steam-pressure as shown by gauge, increased by 14.7.

(*c*) For locomotive boilers : Take the product of (1) the area of the grate-surface in square feet, (2) 11.67, and divide this product by the steam-pressure as shown by gauge, increased by 14.7.

To illustrate the rules, suppose that the steam-pressure in a locomotive boiler is 150 lbs. by gauge ; what is the proper area of opening for the escape of steam by the safety-valve, the grate-surface being 16 square feet ?

The product of 16 and 11.67 is 186.72, and the quotient arising from dividing this by the sum of 150 and 14.7, or 164.7, is about $1\frac{13}{100}$ square inches, which is the required area of opening.

B. *The diameter of valve, in inches, required, to afford the neces-sary area of opening with the given lift.*

(*a*) When the lift of the valve is equal to or less than the depth of the seat : Diminish the required area of opening by the product of (1) the square of the lift, in inches ; (2) the square of the sine of the angle of bevel of the valve ; (3) the cosine of the angle of bevel of the valve, and (4) 3.1416. Divide this difference by the product of (1) the lift in inches ; (2) the sine of the angle of bevel of the valve, and (3) 3.1416.

(*b*) When the lift of the valve is greater than the depth of the seat : Diminish the required area of opening by the product of (1) the square of the depth of seat, in inches ; (2) the square of the sine of the angle of bevel of the valve ; (3) the cosine of the angle of bevel of the valve, and (4) 3.1416. Divide this difference by 3.1416 times the sum of (1) the depth of seat in inches, multiplied by the sine of the angle of bevel of the valve, and (2) the difference between the lift and the depth of seat, in inches.

A table of sines and cosines of angles from 20° to 50° will be found below, and an example is appended in illustration of the rules.

Angle.	Sine.	Cosine.	Angle.	Sine.	Cosine.
20°	.342	.940	36°	.588	.809
21°	.358	.934	37°	.602	.799
22°	.375	.927	38°	.616	.788
23°	.391	.921	39°	.629	.777
24°	.407	.914	40°	.643	.766
25°	.423	.906	41°	.656	.755
26°	.438	.899	42°	.669	.743
27°	.454	.891	43°	.682	.731
28°	.469	.883	44°	.695	.719
29°	.485	.875	45°	.707	.707
30°	.500	.866	46°	.719	.695
31°	.515	.857	47°	.731	.682
32°	.530	.848	48°	.743	.669
33°	.545	.839	49°	.755	.656
34°	.559	.829	50°	.766	.643
35°	.574	.819			

Example.—A safety-valve has a bevel of 33°, a depth of seat of ¼ inch, and is required to give an area of opening of 2 inches, with a lift of ¼ inch. What should be its diameter?

Square of depth of seat..................................0.0625
Square of sine of 33°.............................. ....0.297
                                                      ————
    Product.................................... 0.019
Cosine of 33°...0.839
                                                      ————
    Product...0.016

```
    Product (brought forward)...................... ...0.016
Multiply by.....  ...........   .....................3.1416
    _____
    Product........... .............................0.05
Area of opening.................................  ....2.00 in.
Subtract........  .............................0.05
    _____
Difference is.........................................1.95
Depth of seat.........................................0.25
Sine of 33°............................  .........................0.545
    _____
    Their product......................................0.136
Multiply by..........................................3.1416
    _____
    Product.........  ..............................0.427
Lift................................................0.50
Subtract depth of seat ..............................0.25
    _____
    The difference is.................................0.25
Multiply by.......................................  .........3.1416
    _____
    And the product is..............................0.785
Add...........................................  .........................0.427
    _____
                                                    1.212
```

Now 1.95 divided by 1.212 gives the diameter of valve, 1.61
inches, nearly.   B.

VALVE, SAFETY, Proportions of parts of.—1st. *To find the pressure per square inch at which a given valve will open.* Measure the
following distances horizontally from the fulcrum to (1) the
centre of the valve-stem ; (2) the centre of the weight; (3) the
centre of gravity of the lever, or the point on which it will balance if placed upon a knife-edge. Measure the diameter of the
valve, and determine its area, either from a table or by multiplying the square of the diameter by 0.7854. Find the weight of
(1) the valve ; (2) the lever ; (3) the ball. Multiply (1) the
weight of the ball by its horizontal distance from the fulcrum ;
(2) the weight of the lever by its horizontal distance from the
fulcrum ; (3) the weight of the valve by its horizontal distance
from the fulcrum ; (4) the area of the valve by its horizontal
distance from the fulcrum. Add together the first three products
and divide the sum by the fourth product.

Example.—A given safety-valve has a weight of 50 lbs. 24
inches from the fulcrum, the lever weighs 6 lbs., and its centre of
gravity is 15 inches from the fulcrum ; the weight of the valve
is 2 lbs., and its centre is 4 inches from the fulcrum. The diameter of the valve is 2 inches. At what pressure will the valve
begin to rise ?

```
Square of diameter..................................  4
Multiply by......................................... 0.7854
    _____
Area of valve in square inches..................... 3.1416
```

(1) 50 times 24 is 1200 ; (2) 6 times 15 is 90 ; (3) 2 times 4 is

8 ; (4) 4 times 3.1416 is 12.5664. The sum of (1), (2), and (3) is 1298, which divided by 12.5664 (the fourth product) is 103.03, the pressure in lbs. per square inch at which the valve will open.

2d. *To find where to place the weight on a safety-valve so that it shall open at a given pressure of steam.* Multiply (1) the weight of the lever by the horizontal distance of its centre of gravity from the fulcrum ; (2) the weight of the valve by its horizontal distance from the fulcrum ; (3) the area of the valve by the pressure of steam in lbs. per square inch, and by the horizontal distance of the valve from the fulcrum. Add together the first two products, subtract their sum from the third product, and divide the difference by the weight of the ball.

Example.—The ball of a safety-valve weighs 100 lbs., the lever weighs 10 lbs., the valve weighs 2 lbs., and has a diameter of 3 inches. The distance of the centre of gravity of the lever from the fulcrum is 25 inches, and the distance of the centre of the valve from the fulcrum is 5 inches. How far from the fulcrum must the valve be placed, in order that the lever may open at a pressure of 100 lbs.?

Area of valve, 7.07 square inches.

(1) 10 times 25 is 250 ; (2) 2 times 5 is 10 ; (3) the product of 7.07, 100, and 5, is 3535. Adding together products (1) and (2), we have as their sum 260 ; subtracting this from 3535, the third product, we have 3275. Dividing this difference by 100, the weight of the ball, we have 32.75, or 32¾ inches as the distance from fulcrum to ball.

To find what diameter a safety-valve must have, the other parts being known to open at a given steam-pressure. Multiply (1) the weight of the ball by its horizontal distance from the fulcrum ; (2) the weight of the lever by the horizontal distance of its centre of gravity from the fulcrum ; (3) the weight of the valve by the horizontal distance of its centre from the fulcrum ; (4) the pressure of steam in pounds per square inch by the horizontal distance of the valve from the fulcrum, and by the number 0.7854. Add together the first three products, divide their sum by the fourth product, and take the square root of the quotient.

Example.—Weight of ball, 60 lbs. ; lever, 7 lbs. ; valve, 3 lbs. Distances from fulcrum : ball, 30 inches ; centre of gravity of lever, 16 inches ; centre of valve, 3 inches. Pressure of steam, 70 lbs. per square inch. What should be the diameter of the valve ?

(1) The product of 60 and 30 is 1800 ; (2) the product of 7 and 16 is 112 ; (3) the product of 3 and 3 is 9 ; (4) the product of 70, 3, and 0.7854 is 164.934. The sum of the first three products, 1800, 112, and 9, is 1921. Dividing this sum by 164.934 (the fourth product), we have 11.647 inches. The square root of this number is 3.41+ inches, which by the rule is the required diameter of the valve. **B.**

VALVE, To find the angle of bevel of a.—This is the angle of inclination *f c a*, or *e d b*, to a vertical line. Make the following measurements : (1) greatest diameter, *g h,* of valve, in inches ; (2) least diameter, *a b*, of valve, in inches ; (3) depth, *a k*, of valve, in inches. Divide the difference of the greatest and least diameters by the depth of seat. Find the angle whose tangent is nearest this quotient, in the accompanying table of tangents.

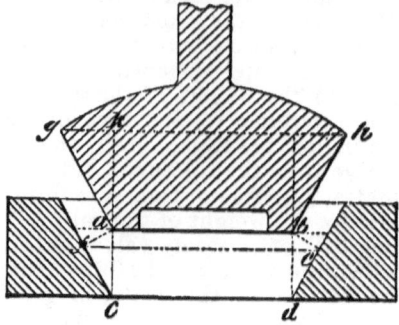

Table of Tangents from 20° to 50°.

Angle.	Tangent.	Angle.	Tangent.
20°	.364	36°	.727
21°	.384	37°	.754
22°	.404	38°	.781
23°	.424	39°	.810
24°	.445	40°	.839
25°	.466	41°	.869
26°	.488	42°	.900
27°	.510	43°	.933
28°	.532	44°	.966
29°	.554	45°	1.000
30°	.577	46°	1.036
31°	.601	47°	1.072
32°	.625	48°	1.111
33°	.649	49°	1.150
34°	.675	50°	1.192
35°	.700		

Example.—The greatest and least diameter of a valve are 4 6-10 and 4 inches, respectively, and the depth is ¼ inch. What is the bevel?

Greatest diameter..................................... 4.6
Least diameter.. 4.

2)0.6

0.5)0.3

Tangent of angle of inclination...................... 0.6
From the table, it appears that the angle corresponding to this is nearly 31°.　　　　　　　　　　　　　　　　　　　B.

VALVE, To find the area of opening, in square inches, of a, due to a given lift.—(*a*) When the lift of the valve is equal or to less than the depth of seat : Find the product of (1) the diameter of the valve, in inches ; (2) the lift, in inches ; (3) the sine of the angle of bevel of the valve, and (4) 3.1416. Add this to the product of (1) the square of the lift, in inches ; (2) the square of the sine of angle of bevel of the valve ; (3) the cosine of the angle of bevel of the valve, and (4) 3.1416.

(b) When the lift of the valve is greater than the depth of seat : Find the product of (1) the diameter of the valve. in inches ; (2) the depth of seat, in inches ; (3) the sine of the angle of bevel of the valve, and (4) 3.1416. Find the product of (1) the square of the depth of seat, in inches ; (2) the square of the sine of the angle of bevel of valve ; (3) the cosine of the angle of bevel of valve, and (4) 3.1416. Find the product of (1) the diameter of the valve, in inches ; (2) the difference between the lift and the depth of seat, in inches, and (3) 3.1416. Take the sum of these three products.

Example.—The diameter of a valve is 4 inches, the bevel is 35°, and the depth of seat ¼ of an inch. What is the area of opening for a lift of ⅛ of an inch ?

The product of 4, 0.25, 0.574 (the sine of 35°), and 3.1416 is 1.8.

The product of the square of 0.25, the square of 0.574, 0.819, (the cosine of 35°), and 3.1416 is 1.85.

The product of 4, 0.125 (the difference between the lift and depth of seat), and 3.1416 is 1.57.

The sum of 1.8, 1.85, and 1.57 is 3.42 square inches, the area of opening required.   B.

BELTS, PULLEYS, AND SHAFTING.

BELT-HOLES, Laying out, through floors.—If a belt is to be carried from a pulley on an overhead shaft to one on any floor above, the distance from centre of lower shaft to ceiling—under side of floor—should be measured and noted ; then the thickness of floor ; next the distance between top of floor and centre of upper shaft. If one pulley or shaft is directly over the other, the size of pulleys and width of belt being known, you have all the data necessary if you measure the distance of one shaft from the wall of building, which is done by dropping a plummet from centre of shaft or diameter of pulley, and measuring to the wall from that point. From these data, whether the two shafts are in the same vertical plane, whether the diameters of the pulleys are equal, and whether the belt is to be carried through one, two, three, or even four floors or not, the intelligent mechanic can lay out a diagram that will enable him to cut his belt-holes accurately. The diagram may be laid out full-size on a swept floor, or, on a reduced scale, on a board or sheet of paper. Measures thus made can easily be transferred to the floor through which the holes are to be made.

BELT-LACING, Eel-skin.—A mill-owner says: "Eel-skins make the best possible strings for lacing belts. One lace will outlast any belt, and will stand wear and hard usage where hooks or any other fastenings fail. Our mill being on the bank of the river, we keep a net set for eels, which, when wanted, are taken out in the morning and skinned, and the skins are stuck on a smooth board. When dry, we cut them in two strings, making the eel-skin, in three hours from the time the fish is taken from the water, travel in a belt."

BELT-LACINGS, Holes for.—The strain on belts is always in the direction of their length, and therefore holes cut for the reception of lacing should be oval (the long diameter in line with the belt). In butting or meeting belts, the crossings of the lacings should be on the outside.

BELT PASSING OVER TWO PULLEYS, To find the length of a.—Measure the distance between the centres of the pulleys, the diameters of the pulleys, and the thickness of the belt. Add the thickness of the belt to the diameter of a pulley, and this gives the effective diameter. Half this is the effective radius, and it is to be noted that the effective radius, or the effective diameter, of a pulley should generally be used in all calculations relating to belts and pulleys. In making such calculations, care must be taken, also, to have all the dimensions in the same unit, feet or inches. In general, it is well to reduce all dimensions to feet.

To illustrate the preceding remarks, suppose that the diameter of a pulley is 10 inches, and that the thickness of the belt passing over it is $\frac{3}{8}$ of an inch. What is the effective radius, in feet? *Ans.* The effective diameter is 10$\frac{3}{8}$ inches ; hence the effective radius is 5$\frac{3}{16}$ inches. 5 inches is 0.417 ft. $\frac{3}{16}$ of an inch is 0.016 ft. Hence 5$\frac{3}{16}$ inches is 0.433 ft.

There are two cases to be considered, one in which the belt is crossed, and the other in which it is open.

To find the length of a crossed belt passing over two pulleys :

(1) Divide the sum of the radii of the two pulleys by the distance between their centres, and find from the table of factors the factor corresponding to this quotient.

(2) Multiply the factor so found by the sum of the radii.

(3) Multiply the sum of the radii by the number 3.1416.

(4) Subtract the square of the sum of the radii from the square of the distance between centres, and take the square root of the remainder. Multiply the quantity so obtained by 2.

(5) Take the sum of the quantities obtained by (2), (3), and (4).

Example.—The radius of one pulley is 42 inches, of the other 36 ; the distance between centres of pulleys is 12 feet, and the thickness of the belt is $\frac{1}{4}$ of an inch ; required, the length of the belt.

The effective radii are 3.51 feet and 3.01 feet.

(1) Sum of radii, 6.520. Distance between centres, 12. Quotient of first quantity divided by second, 0.54. Factor in table corresponding to this quotient, 1.141.

(2) 1.141 multiplied by 6.52, 7.439 +.

(3) 6.541 multiplied by 3.1.416.20.483 +.

(4) Square of distance between centres................144.
Square of sum of radii.. . ... 42.51

Difference................................101.49
Square root of difference, 10.074+. 10.062 multiplied by 2,
20.148.

(5) Sum of 7.439, 20.483, and 20.148, 48.07 feet, or 48 feet and
$\frac{17}{24}$ of an inch, length of belt required.

Table of Factors for Determining the Length of Belts.

Quotient.	Factor.	Quotient.	Factor.	Quotient.	Factor.
0.01	0.020	0.35	0.716	0.69	1.523
0.02	0.040	0.36	0.737	0.70	1.551
0.03	0.060	0.37	0.758	0.71	1.580
0.04	0.080	0.38	0.780	0.72	1.608
0.05	0.100	0.39	0.802	0.73	1.637
0.06	0.120	0.40	0.823	0.74	1.666
0.07	0.140	0.41	0.845	0.75	1.696
0.08	0.161	0.42	0.867	0.76	1.727
0.09	0.180	0.43	0.890	0.77	1.758
0.10	0.201	0.44	0.912	0.78	1.790
0.11	0.220	0.45	0.934	0.79	1.822
0.12	0.241	0.46	0.956	0.80	1.855
0.13	0.261	0.47	0.979	0.81	1.888
0.14	0.281	0.48	1.002	0.82	1.923
0.15	0.301	0.49	1.025	0.83	1.958
-0.16	0 322	0.50	1.047	0.84	1.995
0.17	0.342	0.51	1.070	0.85	2.032
0.18	0.362	0.52	1.094	0.86	2.071
0.19	0.383	0.53	1.118	0.87	2.111
0.20	0.403	0.54	1.141	0.88	2.152
0.21	0.424	0.55	1.165	0.89	2.195
0.22	0.444	0.56	1.189	0.90	2.240
0.23	0.464	0.57	1.214	0.91	2.287
0.24	0.485	0.58	1.238	0.92	2.336
0.25	0.506	0.59	1.262	0.93	2.389
0.26	0.527	0.60	1.287	0.94	2.446
0.27	0.547	0.61	1.312	0.95	2.507
0.28	0.568	0.62	1.338	0.96	2.574
0.29	0.589	0.63	1.364	0.97	2.651
0.30	0.610	0.64	1.389	0.98	2.743
0.31	0.631	0.65	1.415	0.99	2.859
0.32	0.652	0.66	1.443	1.00	3.142
0.33	0.673	0.67	1.469	....	.....
0.34	0.694	0.68	1.496	....	.....

To find the length of an open belt passing over two pulleys :
(1) Divide the difference of the radii by the distance between

centres, and find from the table of factors the factor corresponding to this quotient.

(2) Multiply the factor so found by the difference of the radii.

(3) Multiply the sum of the radii by the number 3.1416.

(4) Subtract the square of the difference of the radii from the
square of the distance between centres, and take the square root
of the remainder. Multiply the quantity so obtained by 2.

(5) Take the sum of the quantities obtained by (2), (3), and (4).
It will be observed that these rules require only simple arithmetical operations.

Example.—Given, diameter of driving-wheel, 36 inches ; of
driven wheel, 9 inches ; distance between centres, 5 feet ; thickness of belt, $\frac{1}{4}$ of an inch ; what is the length of the belt?

Effective radii, 1.505 and 0.380 feet. (1) Difference of radii,
1.125. Distance between centres, 5. Quotient, 0.23. Factor in
table corresponding to this quotient, 0.464. (2) 0.464 multiplied
by 1.125, 0.522. (3) Sum of radii, 1.885. 1.885 multiplied by
3.1416, 5.922.

(4) Square of distance between centres................25.

Square of difference of radii........................ 1.266

Difference................................23.734

Square root of difference, 4.872. 4.872 multiplied by 2, 9.744.
(5) Sum of 0.522, 5.922, and 9.744, 16.188 feet, or 16 feet $2\frac{1}{4}$
inches, length of belt.   **B.**

BELTS, Power transmitted by leather.—By the aid of the accompanying tables, now published for the first time, it will be
easy for any one to ascertain the amount of power that can be
safely transmitted by good leather belts under ordinary circumstances. It is scarcely necessary to add that the power transmitted by a belt in any special case can only be ascertained by
experiment. All that can be done by the most elaborate rules is
to show what power ought to be transmitted if a belt is properly
arranged. The tables and accompanying rules will be useful,
therefore, in calculations of the width of belt required to do a
definite amount of work under given circumstances. With
these preliminary explanations, the use of the tables will be
illustrated.

I. *Other things being equal, the power transmitted by a belt depends upon the arc of contact and the speed of the belt.*

II. *To find the arc of contact between a belt and a pulley, by the
aid of the accompanying table.*

First Method.—Measure the length of the portion of the circumference of the pulley that is in contact with the belt, and
the diameter of the pulley. Divide the first measurement by the
radius of the pulley, which gives the length of the arc of contact
for a circle whose radius is 1. Find the number nearest to this
in the column of the table headed "Length of arc for a radius
of 1," and the required angle will be found in the same horizontal line of the next column, to the left, headed "Arc of contact."

Table for Finding the Arc of Contact of a Belt with a Pulley.

CONSTANT.	Both pulleys, crossed belt, and large pulley, open belt.		Small pulley, open belt.	
	Arc of contact.	Length of arc for a radius of 1.	Arc of contact.	Length of arc for a radius of 1.
0.00	180°	3.142	180°	3.142
.01	181°	3.162	179°	3.122
.02	182°	3.182	178°	3.102
.03	183°	3.202	177°	3.082
.04	185°	3.222	175°	3.062
.05	186°	3.242	174°	3.042
.06	187°	3.262	173°	3.022
.07	188°	3.282	172°	3.002
.08	189°	3.303	171°	2.981
.09	190°	3.322	170°	2.962
.10	192°	3.343	168°	2.941
.11	193°	3.362	167°	2.922
.12	194°	3.383	166°	2.901
.13	195°	3.403	165°	2.881
.14	196°	3.423	164°	2.861
.15	197°	3.443	163°	2.841
.16	198°	3.464	162°	2.820
.17	200°	3.484	160°	2.800
.18	201°	3.504	159°	2.780
.19	202°	3.525	158°	2.759
.20	203°	3.545	157°	2.739
.21	204°	3.566	156°	2.718
.22	205°	3.586	155°	2.698
.23	207°	3.606	153°	2.678
.24	208°	3.627	152°	2.657
.25	209°	3.648	151°	2.636
.26	210°	3.669	150°	2.615
.27	211°	3.689	149°	2.595
.28	213°	3.710	147°	2.574
.29	214°	3.731	146°	2.553
.30	215°	3.752	145°	2.532
.31	216°	3.773	144°	2.511
.32	217°	3.794	143°	2.490
.33	219°	3.815	141°	2.469
.34	220°	3.836	140°	2.448
.35	221°	3.858	139°	2.426
.36	222°	3.879	138°	2.405
.37	223°	3.900	137°	2.384
.38	225°	3.922	135°	2.362
.39	226°	3.944	134°	2.340
.40	227°	3.965	133°	2.319
.41	228°	3.987	132°	2.297
.42	230°	4.009	130°	2.275
.43	231°	4.032	129°	2.252
.44	232°	4.054	128°	2.230
.45	234°	4.076	126°	2.208
.46	235°	4.098	125°	2.186
.47	236°	4.121	124°	2.163
.48	237°	4.144	123°	2.140
.49	239°	4.167	121°	2.117
.50	240°	4.189	120°	2.095

Table for Finding the Arc of Contact of a Belt with a Pulley.
(Continued.)

CONSTANT.	Both pulleys, crossed belt, and large pulley, open belt.		Small pulley, open belt.	
	Arc of contact.	Length of arc for a radius of 1.	Arc of contact.	Length of arc for a radius of 1.
.51	241°	4.212	119°	2.072
.52	243°	4.236	117°	2.048
.53	244°	4.260	116°	2.024
.54	245°	4.283	115°	2.001
.55	247°	4.307	113°	1.977
.56	248°	4.331	112°	1.953
.57	250°	4.356	110°	1.928
.58	251°	4.380	109°	1.904
.59	252°	4.404	108°	1.880
.60	254°	4.429	106°	1.855
.61	255°	4.454	105°	1.830
.62	257°	4.480	103°	1.804
.63	258°	4.506	102°	1.778
.64	260°	4.531	100°	1.753
.65	261°	4.557	99°	1.727
.66	263°	4.585	97°	1.699
.67	264°	4.611	96°	1.673
.68	266°	4.638	94°	1.646
.69	267°	4.665	93°	1.619
.70	269°	4.693	91°	1.591
.71	271°	4.722	89°	1.562
.72	272°	4.750	88°	1.534
.73	274°	4.779	86°	1.505
.74	275°	4.808	85°	1.476
.75	277°	4.838	83°	1.446
.76	279°	4.869	81°	1.415
.77	281°	4.900	79°	1.384
.78	283°	4.932	77°	1.352
.79	284°	4.964	76°	1.320
.80	286°	4.997	74°	1.287
.81	288°	5.030	72°	1.254
.82	290°	5.065	70°	1.219
.83	292°	5.100	68°	1.184
.84	294°	5.137	66°	1.147
.85	296°	5.174	64°	1.110
.86	299°	5.213	61°	1.071
.87	301°	5.253	59°	1.031
.88	303°	5.294	57°	0.990
.89	306°	5.337	54°	0.947
.90	308°	5.382	52°	0.902
.91	311°	5.429	49°	0.855
.92	314°	5.478	46°	0.806
.93	317°	5.531	43°	0.758
.94	320°	5.588	40°	0.696
.95	324°	5.649	36°	0.635
.96	328°	5.716	32°	0.568
.97	332°	5.793	28°	0.491
.98	337°	5.885	23°	0.399
.99	344°	6.001	16°	0.283
1.00	360°	6.284	0°	0.000

Example.—Suppose the length of the circumference of a pulley in contact with a belt is 8½ feet, and the diameter of the pulley is 4 feet.   The quotient arising from dividing 8½ by 2 (the radius of the pulley) is 4.25, and the number in the table nearest to this is 4.26, showing that the required arc of contact is about 53°.

[It is to be noted that, in calculations of this kind, the effective radius of the pulley should be used (see page 110), and all dimensions must be referred to the same unit of measurement.]

Second Method.—Measure the effective diameters of both pulleys, and the distance between their centres.   There are two cases to be considered :

(a) *To find the arc of contact for a crossed belt.*

Divide the sum of the radii of the two pulleys by the distance tween their centres ; find in the column of constants the nearest number to the quotient, and pick out the corresponding angle.

Example.—Diameter of driven pulley, 20 inches ; of driving pulley, 24 ; distance between centres, 8 feet.   What is the arc of contact on each pulley of a crossed belt passing over them ? Sum of radii, 1.8333 feet.   1.8333 divided by 8 is 0.23, nearly.   From the table, it appears that the angle required is 207°.

(b) *To find the arc of contact for an open belt.*

Divide the difference of the radii of the two pulleys by the distance between their centres, and find the angles corresponding to the constant nearest to the quotient, in the table.

Example.—In the case of an open belt passing over two pulleys, the following dimensions are given : Diameter of driving pulley is 5.25 feet ; diameter of driven pulley is 3.5 feet ; distance between centres is 9 feet.   The difference of radii (0.875) divided by 9 is 0.097+.   Nearest constant in table, 0.1, corresponding to an angle of contact of 192° on the driving, and 168° on the driven pulley.

[These rules are founded on the assumption that the belt is drawn perfectly tight between the pulleys.   Where there is much deviation from this, in practice, it is better to employ the first method.]

III. *To find the speed of a belt, in feet, per minute.*

Multiply the diameter of either pulley, in feet, by 3.1416 times the number of revolutions that it makes per minute.

Example.—A belt passes over a pulley that is 3 feet in diameter, and makes 200 revolutions a minute.   The speed of the belt is the product of 3, 3.1416, and 200, or about 1885 feet per minute.

IV. *To find the power that can be safely transmitted by a good leather belt of given width, passing over smooth iron pulleys, and running at a given speed, the arc of contact being also given.*

This is determined by means of the second table.   Find the horse-power for a belt one inch in width, for the nearest arc of contact in the table, and the nearest speed of belt, and multiply this by the width of the belt.   If the belt is open, and the pulleys have different diameters, take the angle of contact made by the belt with the smaller pulley.

Horse-Power Transmitted by a Leather Belt One Inch Wide.

SPEED OF BELT, IN FEET, PER MINUTE	Horse-power transmitted for arc of contact of											
	10°	20°	30°	40°	50°	60°	70°	80°	90°	100°	110°	120°
100	0.014	0.028	0.040	0.052	0.062	0.072	0.082	0.090	0.098	0.106	0.112	0.119
200	0.029	0.056	0.080	0.103	0.125	0.145	0.163	0.180	0.196	0.211	0.225	0.237
300	0.043	0.083	0.120	0.155	0.187	0.217	0.245	0.270	0.294	0.316	0.337	0.356
400	0.058	0.111	0.161	0.207	0.249	0.289	0.326	0.360	0.392	0.422	0.449	0.475
500	0.072	0.139	0.201	0.258	0.312	0.361	0.408	0.451	0.470	0.527	0.562	0.594
600	0.086	0.166	0.241	0.310	0.374	0.434	0.489	0.541	0.588	0.633	0.674	0.712
700	0.101	0.194	0.281	0.361	0.436	0.506	0.571	0.631	0.686	0.738	0.786	0.831
800	0.115	0.222	0.321	0.413	0.499	0.578	0.652	0.721	0.785	0.844	0.899	0.950
900	0.129	0.250	0.361	0.465	0.561	0.651	0.734	0.811	0.883	0.949	1.011	1.068
1000	0.144	0.277	0.401	0.516	0.624	0.723	0.815	0.901	0.981	1.055	1.123	1.187
2000	0.288	0.555	0.803	1.033	1.247	1.446	1.630	1.802	1.961	2.109	2.247	2.344
3000	0.431	0.832	1.204	1.549	1.871	2.169	2.446	2.703	2.942	3.164	3.370	3.561
4000	0.575	1.109	1.605	2.065	2.494	2.892	3.261	3.604	3.922	4.218	4.493	4.748
5000	0.719	1.386	2.007	2.581	3.118	3.614	4.076	4.505	4.903	5.273	5.616	5.936
6000	0.863	1.664	2.408	3.098	3.741	4.337	4.891	5.406	5.883	6.327	6.740	7.123

Horse-Power Transmitted by a Leather Belt One Inch Wide.—Continued.

SPEED OF BELT, IN FEET, PER MINUTE.	Horse-power transmitted for arc of contact of											
	130°	140°	150°	160°	170°	180°	190°	200°	210°	220°	230°	240°
100	0.125	0.130	0.135	0.140	0.144	0.149	0.152	0.156	0.159	0.162	0.165	0.168
200	0.249	0.260	0.270	0.280	0.289	0.297	0.305	0.312	0.318	0.324	0.330	0.335
300	0.374	0.390	0.405	0.420	0.433	0.446	0.457	0.468	0.477	0.487	0.495	0.503
400	0.499	0.521	0.540	0.560	0.578	0.594	0.609	0.624	0.637	0.649	0.660	0.671
500	0.623	0.651	0.675	0.700	0.722	0.743	0.762	0.779	0.796	0.811	0.825	0.838
600	0.748	0.781	0.810	0.840	0.867	0.891	0.914	0.935	0.955	0.973	0.990	1.006
700	0.873	0.911	0.935	0.980	1.011	1.040	1.066	1.091	1.114	1.135	1.155	1.174
800	0.997	1.041	1.080	1.120	1.155	1.188	1.219	1.247	1.273	1.298	1.320	1.341
900	1.122	1.171	1.215	1.260	1.300	1.337	1.371	1.403	1.432	1.460	1.485	1.509
1000	1.246	1.301	1.350	1.400	1.444	1.485	1.523	1.559	1.592	1.622	1.650	1.677
2000	2.493	2.603	2.699	2.800	2.888	2.970	3.047	3.117	3.183	3.244	3.301	3.353
3000	3.739	3.904	4.048	4.200	4.333	4.456	4.570	4.676	4.774	4.866	4.951	5.030
4000	4.986	5.206	5.398	5.600	5.777	5.941	6.093	6.235	6.366	6.488	6.601	6.707
5000	6.232	6.507	6.747	7.000	7.221	7.426	7.616	7.793	7.957	8.110	8.252	8.383
6000	7.478	7.809	8.097	8.401	8.665	8.911	9.140	9.352	9.549	9.731	9.902	10.060

Horse-Power Transmitted by a Leather Belt One Inch Wide.—Continued.

Horse-power transmitted for arc of contact of

SPEED OF BELT, IN FEET, PER MINUTE.	250°	260°	270°	280°	290°	300°	310°	320°	330°	340°	350°	360°
100	0.170	0.172	0.175	0.176	0.178	0.180	0.182	0.183	0.184	0.186	0.187	0.188
200	0.340	0.345	0.349	0.353	0.356	0.360	0.363	0.366	0.369	0.371	0.374	0.376
300	0.510	0.517	0.524	0.529	0.535	0.540	0.545	0.549	0.553	0.557	0.560	0.564
400	0.680	0.690	0.698	0.706	0.713	0.720	0.726	0.732	0.737	0.742	0.747	0.751
500	0.851	0.862	0.872	0.882	0.891	0.900	0.908	0.915	0.922	0.928	0.934	0.939
600	1.021	1.034	1.047	1.059	1.069	1.080	1.089	1.098	1.106	1.114	1.121	1.127
700	1.191	1.207	1.221	1.235	1.248	1.260	1.271	1.281	1.290	1.299	1.307	1.315
800	1.361	1.379	1.396	1.412	1.426	1.440	1.452	1.464	1.475	1.485	1.494	1.503
900	1.531	1.551	1.570	1.588	1.604	1.620	1.634	1.647	1.659	1.670	1.681	1.691
1000	1.701	1.724	1.745	1.765	1.782	1.800	1.815	1.830	1.843	1.856	1.868	1.879
2000	3.402	3.448	3.490	3.529	3.564	3.599	3.631	3.660	3.687	3.712	3.735	3.757
3000	5.103	5.171	5.235	5.293	5.346	5.399	5.446	5.490	5.530	5.568	5.603	5.636
4000	6.804	6.896	6.980	7.058	7.129	7.198	7.261	7.319	7.374	7.424	7.471	7.514
5000	8.505	8.619	8.724	8.822	8.911	8.998	9.076	9.149	9.217	9.280	9.338	9.393
6000	10.206	10.343	10.469	10.587	10.693	10.798	10.892	10.979	11.060	11.136	11.206	11.271

Example.—What horse-power can be transmitted by a leather belt 10 inches wide, making an angle of contact of 80° with the smaller of the two pulleys over which it passes, and having a speed of 2400 feet per minute? *Ans.*—Horse-power for a belt 1 inch wide and 2000 feet speed is 1.802 ; horse-power for a belt 1 inch wide and 400 feet speed is 0.360 ; therefore, by addition, horse-power of 1 inch belt for 2400 feet speed is 2.162; and for 10 horse-power, 10 times 2.162, or 21.62.

V. *To find the width of belt necessary to transmit a given amount of power for a given arc of contact and given speed in feet per minute.*

Find the power transmitted by a belt 1 inch wide, and divide the given power by this amount.

Example.—An open belt passes over two pulleys having diameters of 4 and 6 feet respectively, and the former makes 300 revolutions a minute. The distance between the centres of the pulleys is 15 feet. What should be the width of a belt to transmit 50 horse-power under these circumstances? Speed of belt in feet per minute, 3770 ; arc of contact of belt with smaller pulley, 172° ; horse-power transmitted by a belt 1 inch wide, under conditions in table nearest to those determined above (*i.e.*, for arc of contact of 170°, and speed of 3800), 5.488. Required width of belt, 50 divided by 5.488, or a little over 9 inches.

VI. In the use of a leather belt, it is best to run it with the grain side next to the pulleys, or in exactly the opposite way from that in which the hide was worn by the animal that was the original proprietor of the leather.

VII. Lace-leather is better than hooks for fastening the ends of a belt together ; and a still better method, after a belt has become sufficiently stretched by use, is to rivet the ends together with long laps. In lacing a belt of any considerable size, make two rows of holes in each belt end, and put in double lacing.

VIII. A belt that is made of good material, and is of ample size, will last for many years, if kept clean, and prevented from becoming dry and hard by the use of neat's-foot oil. It is poor economy to buy a belt whose chief recommendation is its small first cost. It is also a bad plan to use a belt that is just sufficient to transmit the power when very tightly strained. B.

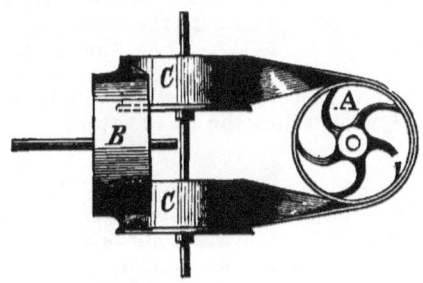

CORNER-TURNED BELT.

BELTS, CORNER-TURNED, Device for.—The two shafts placed at right angles, as shown in the engraving, carry the belt from A or

B, passing around two flanged pulleys or guides, C, turning loosely on a fixed upright shaft, and sustained in position by a collar
under the hub of each. It is possible to run pulleys by this device which not only have varying diameters, but the shafts of
which are on different levels ; but the results are not so good,
owing to unequal strain on the belt. It is better to confine this
method to shafts on the same level, and to pulleys of equal diameter, and the useful limit of angle of shafts is that of 45°, or
less. A greater or more obtuse angle is better run by means of
guides on two uprights.

BELTS, COUPLING.—In Fig. 1, A, B, C, and D are pieces of No.
16 sheet-iron, riveted to the ends of the belt ; E F are hooks,
shown in the natural size in Fig. 2, riveted to B. After the belt

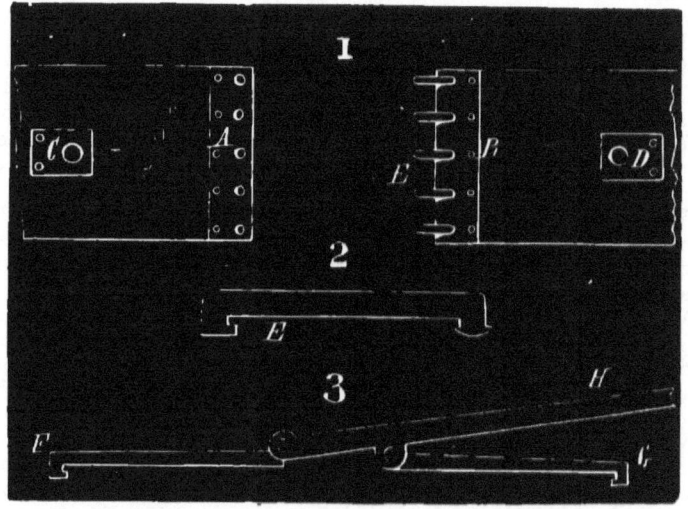

COUPLING BELTS.

is laid over the pulleys, the hooks, F and G, of the lever, shown
in Fig. 3, are placed in the holes at C and D. Now the two ends
of the belt are drawn together by the lever, H, and the hooks,
E, are put in their places at A. Then the lever is taken out,
leaving the joint finished. By this method, two men can set and
couple a belt in the least possible time, obtaining an effectual
joint, which will never allow the belt to run out of true, or to reverse.

BELTS, Increasing the conveying force of.—Adding to the width
of a belt and of the faces of the pulleys increases immensely the
power of conveying force. A wide belt is always better than a
narrow one strained to its utmost capacity.

BELTS, Mending.—Lay the two ends of the belt exactly even,
with the insides together, and punch one straight row of holes
across the end, driving the punch through both pieces so that the

holes may correspond. Now take your lace, pointed at both ends, and pass the points in opposite directions through the first hole, still keeping the two ends of the belt together as when punched, and draw the loop tight, observing to keep the ends of equal length. Pass the points through the second hole, and so proceed to the last ; then tie the ends over the edge of the belt, and the job is done. A belt can thus be mended in half the time and with half the length of lacing required in the usual way ; and when the belt is subjected to heavy strains or slipping, it will wear ten times as long, as the lace never touches the pulley-faces. Of course the plan is not applicable when both sides of a belt run over pulleys, nor when the projecting ends would strike any thing in their track.

BELTS, Oiling.—The best mode of oiling a belt is to take it from the pulleys, and immerse it in a warm solution of tallow and oil ; after allowing it to remain a few moments, the belt should be immersed in water heated to 100° Fahr., and instantly removed. This will drive the oil and tallow all in, and at the same time properly temper the leather.

BELTS, RUBBER, To prevent, slipping on pulleys.—Chalk the pulley when slipping occurs. The presumption is, however, that the belt is either too narrow or too loose.

BELTS, Splicing large.—Cut your belt perfectly square on the ends and to the proper length ; then cut a piece of belt of the same width and thickness, about 3 feet long. Bring the ends of the belt together, and put the short piece on the back of the joint, or outside, and bolt the belt and piece together with what are known as elevator-bolts, used for fastening the buckets to elevator-bands. The tools required are a brace and bit to bore the holes, and a small pair of blacksmith's tongs to tighten up the nuts with.

BELTS, Testing leather for.—A cutting of the material about 0.03 of an inch in thickness is placed in strong vinegar. If the leather has been thoroughly acted upon by the tanning, and is hence of good quality, it will remain, for months even, immersed without alteration, simply becoming a little darker in color. But, on the contrary, if not well impregnated by the tannin, the fibres will quickly swell, and, after a short period, become transformed into a gelatinous mass.

BELTS, To lay out quarter-twist.—To make holes through floors for the belts, lay out on a floor with chalk-line and train two views of the pulley, or by scale on paper as shown in the annexed diagram. B is the belt running in the direction of the arrow on the lower pulley, and C is the belt running in the opposite direction. Therefore, drop a plumb-line, representing the perpendiculars, B and C, and draw the diagonals governed by the diameters of the pulleys, marking the distances $a b$ and $c d$ on the floor. Now, drop a plumb-line from each side of the centre of face of upper pulley to the floor, and from one point, c, thus found, lay off the distance *a b*, in a line parallel with the upper shaft, and from the point *a* in the distance, *c d*, parallel

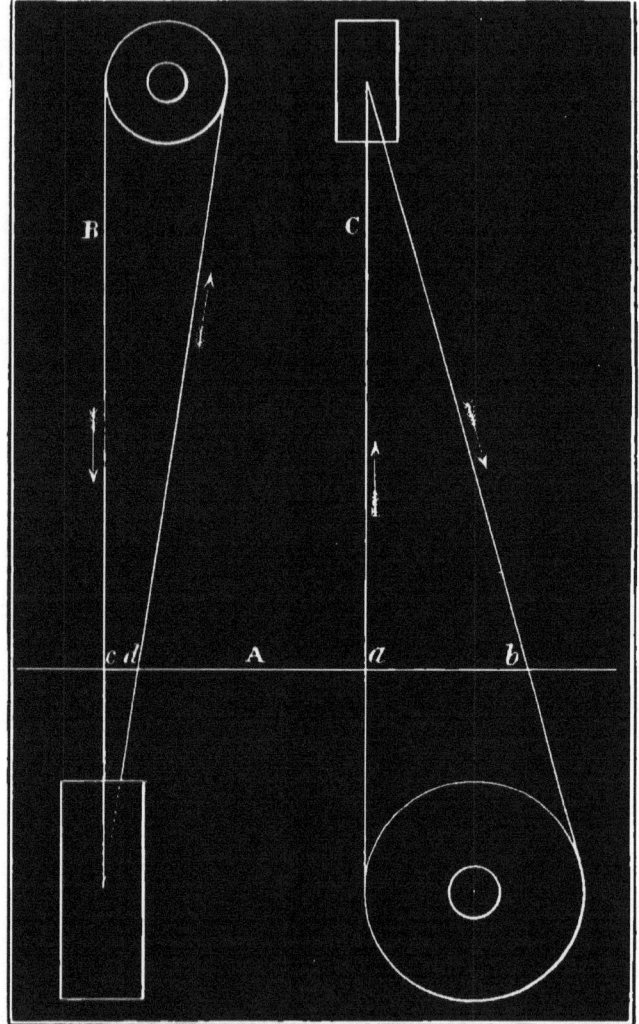

LAYING OUT QUARTER-TWIST BELTS THROUGH FLOORS.

with the lower shaft. These points are the places at which the holes should be cut.

FRICTION AND LUBRICANTS.—Whenever one surface moves upon another, the rough and projecting points of the two surfaces (which always exist, even in the smoothest surfaces) oppose resistance to the motion, and this resistance is called friction.

The coefficient of friction is a quantity expressing the ratio of the friction to the pressure. For instance, if the resistance to moving one piece of metal on another is one fifth of the weight of the moving body, the coefficient of friction in this case is one fifth, or 0.2. Hence, knowing the coefficient of friction, in any given instance, and the weight of the body causing the resistance, the amount of friction is found by multiplying these two quantities together.

The work due to or lost on account of friction, in any given time, is found by multiplying the amount of friction of the moving body by the space passed through in the given time. It is customary to estimate the amount of friction in pounds, to make the given time one minute, and to measure the distance passed through in that time in feet. The result obtained will then express the number of foot-pounds of work performed per minute in overcoming friction, and this can readily be reduced to horse-power, or any other desired unit of work. It is important to maintain the distinction between the amount of friction and the work of friction.

The experiments of Coulomb and Morin have demonstrated the following facts in regard to friction :

That it is proportional to the pressure.

With some limitations, that it is independent of the area of the surface pressed, and independent of the velocity of motion.

The limitations are, that the pressure should not be so great as to abrade or wear away the surface rapidly, in which case the friction does not follow the laws enunciated above ; also, that the velocity of motion shall not be so great as to expel the lubricant. It is found, for instance, in the case of the journals of car-axles, that they require to be enlarged as the speed increases, in order to prevent the expulsion of the lubricant. The actual bearing surface of a journal is usually considered to be the projected area of that journal, or the product of the length multiplied by the diameter. For instance, if a journal is 4 inches in diameter and 7 inches long, the bearing surface is 28 square inches.

The pressure per square inch on the bearing surface should not exceed the following limits :

Velocity of periphery of journal.	Limiting pressure per sq. in. of bearing surface.
1 foot per second	.382 lbs.
2¼ feet per second....................................	.224 "
5 feet per second....................................	.140 "

It is well known that one of the most common expedients for reducing the friction between two rubbing surfaces is to interpose some lubricant, which seems to form a coating to the projecting points, making the whole surface more continuous, and thus lessening the resistance. At very low pressures and velocities the viscosity of the lubricant occasionally causes the resistance to be increased instead of lessened, but in general the effect of an unguent is to decrease the friction in quite a large ratio. Careful experiments have been made with regard to the friction between two surfaces when they were perfectly dry and clean, and when

different lubricants were used.   In this manner it has been found
that good oil, such as olive-oil, is one of the best lubricants ; that
lard is better than tallow, and that the use of water, instead of
lessening the friction, generally increases it.   Experiments upon
the manner of applying the lubricant show that there is a great
advantage in a continual application, so as to keep a film constantly interposed between the rubbing surfaces, over the case in
which the surfaces are merely kept slightly greasy.   Below are
given mean values for the coefficient of friction, in cases arising
from the sliding of one plane surface upon another, the surfaces
being supposed to be true, and, in common language, smooth :

NATURE OF THE SLIDING SURFACE.	Smooth surfaces without lubrication...	WELL LUBRICATED WITH					
		Water........	Soap........	Olive-oil........	Tallow........	Lard........	Lard and plumbago........
Wood on wood..........	0.38	.....	0.144	.....	0.071	0.066	.....
Wood on metal..........	0.41	.....	0.20	0.064	0.079	0.076	.....
Metal on metal	0.18	0.311	0.197	0.071	0.092	0.075	0.070

In the case of journals, the coefficient of friction is generally
much less than for plane surfaces.   Mean values of this coefficient, both for wood and metals, vary from 0.15, when the journal is only slightly unctuous, to 0.05, when there is a continual
supply of the lubricant.   In regard to journal-friction, the amount
is independent of the diameter of the journal, but the work required to overcome friction will of course be greater with a large
than with a small journal, because the distance passed through
by the periphery of the journal in a given time will be greater in
the former case.   B.

BELTS, Testing vulcanized rubber for.—These trials consist in
examining the comparative degrees of elasticity and tenacity.
The manner in which they are conducted in the French navy appears practical and easily followed.   The first test consists in cutting from the sheets samples, which are left in a steam-boiler
under a pressure of 5 atmospheres for 48 hours.   At the end of
this time, the pieces should not have lost their elasticity.   The
specimens may then be placed on the grating of a valve-box, under a pressure from above of 85.5 lbs. per square inch, and should
withstand 9100 strokes at the rate of 100 per minute.   Specimens
not boiled should withstand 17,100 strokes.   Thongs of rubber
boiled, and having a section 0.6 inch square and a length of 8
inches, fixed between supports and elongated 3.9 inches, should
resist, without breaking, a further elongation of 8 inches, re-

peated 22 times a minute for 24 hours. Thongs not boiled, under the same conditions, should resist for 100 hours. These extra elongations may be easily made by a wheel, to the periphery of which one end of the thong is fastened, while the other extremity may be attached to a support. By turning the wheel, any determined elongation may be given at the rate of from 20 to 25 times per minute. Under the above conditions, bands of first quality rubber, perfectly pure and well vulcanized, break after 180 or 200 elongations of 8 times the initial length. Bands cut from pure rubber, but of secondary quality, break after 50 or 60 elongations. Inferior caoutchouc, containing mineral matters or residue of old vulcanized rubber, gives no results at all.

M. Ogier has investigated the properties of rubber belts made of repeated layers of cloth covered with prepared rubber. Through the adhesive nature of the caoutchouc, the superposed tissues form, after vulcanization, a homogeneous substance, comparable, in M. Ogier's opinion, to the best curried leather. His experiments, in order to obtain the coefficient of friction of these belts on cast-iron pulleys, give us results varying from 0.42 to 0.84, as against the coefficient for leather, 0.28. The minimum value corresponds to canvas and rubber belts without an exterior rubber coating. On pulleys of various forms, the maximum value of the coefficient of friction was found on those slightly convex and presenting a roughly turned surface, this result being inverse to that obtained with leather belts. Similarly the presence of fatty bodies has an opposite action on the cloth and rubber belts to that which it has on leather. On covering the former with a light varnish of half olive-oil and half tallow, the adhesion was found to be considerably augmented. (1) The resistance to traction of rubber and canvas belts per square millimeter (0.0009 square inch) of section, is at least equal to that of leather belts. (2) This resistance per square millimeter is independent of dimensions—length, breadth, or thickness. Such is not the case with leather belts, and therefore preference should be given to rubber belting whenever the conditions of the power to be transmitted necessitate the employment of very long, very wide, and very thick belts. (3) From two trials it appears that the external covering of caoutchouc adds nothing to the resistance, and hence it is advantageous to use covered belts which, at equal weights and prices, give a superior resistance. (4) Under the same weight, the elastic elongation of leather belts is double that of rubber ones. The permanent elongation, under a change of 0.55 pound per square millimeter, reached 2 per cent in the former and nothing in the latter.

BELTS, To prevent gnawing of, by rats.—Anoint with castor-oil.

BELT-TIGHTENER's, To place.—The loss of power occasioned by the use of a tightener, is the power required to bend the belt under that pulley and to drive the pulley. By placing the tightener near the smaller pulley of two of unequal size, there is a greater loss than when it is close to the larger, since the belt requires to be more bent in the former case. The best place, therefore, to put a tightener, is as close to the larger pulley as it can be arranged to have it work satisfactorily.

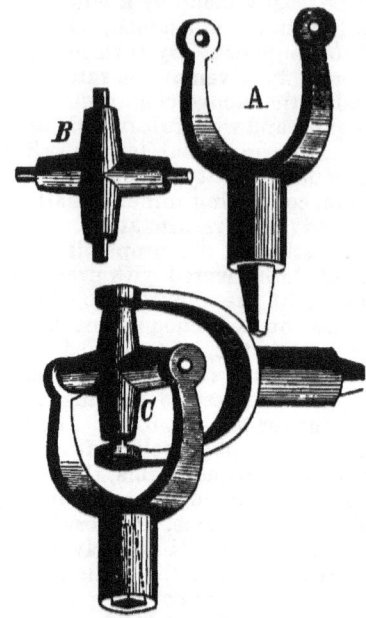

UNIVERSAL POINT FOR BORING-TOOL.

HANGERS, Securing.—If it should be required to place a hanger between flooring-beams, the floor to which it is attached should be strengthened with a generous piece of plank. For securing hangers, lag-screws are superior to bolts with nuts, where there is sufficient thickness of wood. A wooden straight-edge, reaching from one bearing to another, is better for leveling hangers and boxes than a twine, which will sag more or less. Some use short cylinders of iron turned to fit the box, and having a central hole drilled longitudinally through them. This is an excellent plan, as the eye may sight through, or a string be passed through to determine the level. Where holes are to be bored through the floor close to a wall, post, or other vertical obstruction, a handy tool, similar to that shown in the cut, comes into play. It is easily forged, and need not be finished with the elegance of contour shown. A is one of the yokes, and B the cross; they are seen united at C. The shank of one yoke has a tapering square hole to receive a bit or auger, and the other is a tapering square shank to fit a stock of the bit-brace. The device is "a universal joint," and can be readily worked at an angle of 45°.

HOT-BEARING ALARM.—A cylindrical box, A, is provided with a perforated bottom, B, and placed directly over the journal. The box is filled with a prepared grease which melts at a certain temperature, to which it must be raised by the shaft becoming hot. As the compound liquifies and escapes through the perforations, a disk, C, which rests thereon, descends, thereby tilting the lever, D, and so making contact between the plates, E and F. The latter are connected by an electric circuit with a bell which

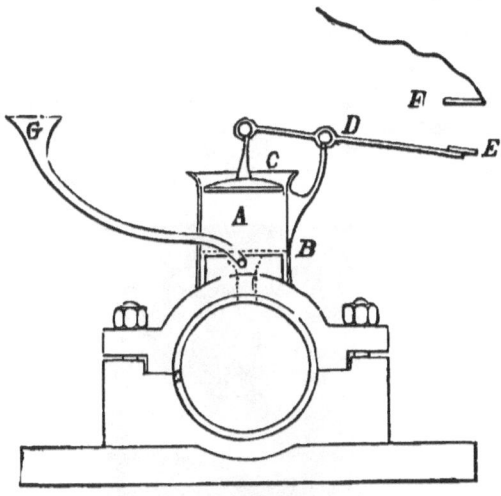

HOT-BEARING ALARM.

sounds when the current is established. The pipe, G, serves for the ordinary lubrication of the journal. It is suggested that this device might be profitably used upon journals not readily accessible.

PULLEYS, Balancing.—Swing the pulley on arbors between lathe-centres, and note the position as determined by gravity. On

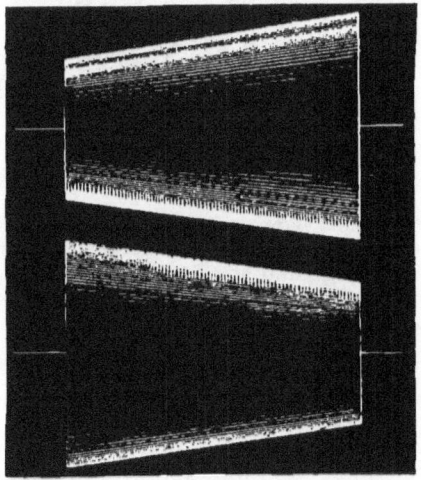

CONE PULLEYS, FIG. 1.

the top side, drill and tap two holes, in which seat machine screws with flat heads, the shanks projecting through from the face or

outer side.   Then, by securing pieces of iron as weights to this
point until the pulley is balanced, the amount necessary to
balance the pulley is found.   This amount of lead is then melted
and cast in a mould formed by clay.   The screws serve to hold
the lead in place.

PULLEYS, To design cone.—The following rules will enable any
one who understands arithmetical operations to make the calculations necessary for designing a set of cone pulleys in such a
manner that the belt can be shifted from one pair to another, and
be equally tight in every position.   There are six cases to be considered.

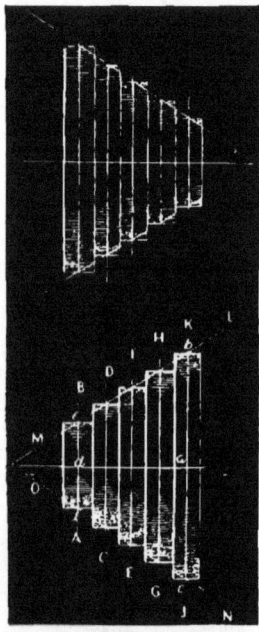

CONE PULLEYS, FIG. 2.

CASE 1.—*Crossed belt passing over two continuous cones.* (Fig. 1.)
—In this case, it is only necessary to use two similar conical
drums, with their large and small ends turned opposite ways.

CASE 2.—*Crossed belt passing over two stepped cones that are
equal and opposite.* (Fig. 2.)   Draw vertical lines, A B, C D, etc.,
to the axes of the pulleys, at a distance apart equal to the face of
a pulley.   Lay off, on each side of the axis, distances, a b, a c, equal
to the radius of the largest pulley, and d e, d f, equal to the radius
of the smallest pulley.   Draw a straight line, L M, through b and
e, and N O through c and f.   The points in which these lines cut
the verticals determine the radii of the intermediate pulleys.

CASE 3.—*Crossed belt passing over any two stepped cones.*—Assume values for the radii of one driving-pulley and the corresponding driven pulley.   Then, for any assumed radius of a

second driving-pulley, the radius of the driven pulley must have such a value that the sum of these two radii is equal to the sum of the first two. The same must be true for every pair of pulleys in the two stepped cones.

Example.—Suppose the radius of the first driving-pulley is 15 inches, and of the first driven pulley 5 inches. Now, if there are five steps in the driving-cone, having radii of 15, 12, 9, 6, and 3 inches respectively, the corresponding steps of the driven cone will have radii of 5, 8, 11, 14, and 17 inches, since the sum of the radii of each pair of pulleys must be equal to the sum of the radii of the first pair, or 20 inches. It will be evident from the foregoing that, in the case of crossed belts, the construction of cone-pulleys is very simple, since it is only necessary to observe the directions given above, no matter what the distance between the centres of driving and driven pulley may be.

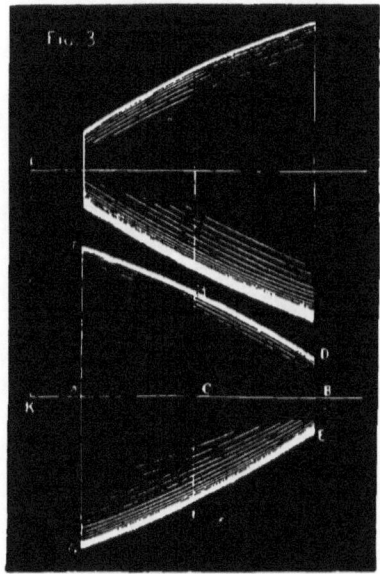

CONE PULLEYS, FIG. 3.

CASE 4.—*Open belt passing over two continuous pulleys.* (Fig. 3.) For this case equal and similar conoids must be used. Assume the largest radius, A F, and the smallest, B D, and calculate, by the rule on page 111, the length of belt required for pulleys with the given radii, the distance, K L, between their centres being given. Then the middle radius, C H, is found by the following rule :

Subtract twice the distance between centres from the length of the belt, and divide the difference by the number 6.2832.

Having found the middle radius, draw circular arcs through the points F H D and G I E, thus determining the section of the conoid.

Example.—Suppose that the largest radius is 24 inches, the smallest 6 inches, and the distance between centres of conoids 3 feet. What should be the middle radius?

First find the length of belt: 2 diminished by 0.5 equals 1.5. This divided by 3 equals 0.5, and the corresponding number in table of factors, page 111, is 1.047—(1). 1.047 multiplied by 1.5 equals 1.571—(2). 2 added to 0.5 equals 2.5, which multiplied by 3.1416 equals 7.854—(3). 3 multiplied by 3 equals 9, which less 2.25 equals 6.75. 1.5 multiplied by 1.5 equals 2.25. The square root of 6.75 is 2.6, which multiplied by 2 equals 5.2—(4). The sum of 5.2 and 1.571 and 7.854 equals 14.625, which is the length of belt. Then find the middle radius by the preceding rule: 3 multiplied by 2 equals 6. 14.625 less 6 equals 8.625, which divided by 6.2832 equals 1.373 feet, or about 16½ inches, middle radius required.

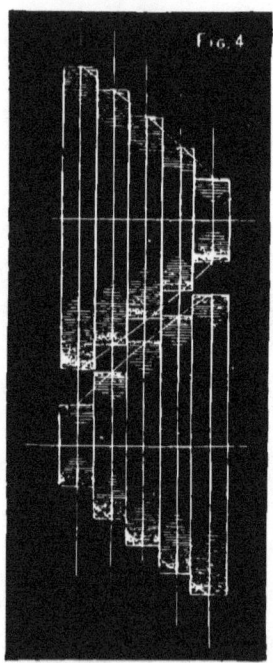

CONE PULLEYS, FIG. 4.

CASE 5.—*Open belt passing over two stepped cones that are equal and opposite.*—(Fig. 4.)—The construction will be evident from the figure, it only being necessary to form two continuous conoids, as explained above, and divide them into the required number of steps.

CASE 6.—*Open belt passing over any two stepped cones.*—The rules for this case, originally demonstrated by J. B. Henek, are presented below in a simplified form. First assume the radii of one driving-pulley and the corresponding driven pulley, measure the distance between their centres, and find the length of belt required. Then assume values for the radii of the successive pulleys on the driving-cone, and calculate the values of the corresponding radii on the driven cone by the following rules : I. Having assumed the value of one radius, it is first necessary to ascertain whether the one to be calculated is larger or smaller. (1) Multiply the assumed radius by the number 3.1416, and increase the product by the distance between the centres of the pulleys. (2) If the quantity obtained by (1) is greater than half the length of the belt, the assumed radius is greater than the one to be determined. (3) If the quantity obtained by (1) is less than half the length of the belt, the assumed radius is less than the one to be determined. II. *When the assumed radius is the greater of the two, to find the other one.* The distance between centres and the length of belt are supposed to be given. (1) Multiply the assumed radius by the number 6.2832, subtract this product from the length of the belt, and divide the remainder by the distance between centres. (2) Add the quantity obtained by (1) to the · number 0.4674, and extract the square root of the sum. (3) Subtract the quantity obtained by (2) from the number 1.5708, and multiply the difference by the distance between centres. (4) Subtract the quantity obtained by (3) from the assumed radius ; the remainder will be the required radius. III. *When the assumed radius is the smaller of the two, to find the other one.* (1) Same as (1) of preceding rule. (2) Same as (2) of preceding rule. (3) Subtract the number 1.5708 from the quantity obtained by (2), and multiply the difference by the distance between centres. (4) Add the quantity obtained by (3) to the assumed radius ; the sum will be the required radius.

Example.—The first driving-pulley of a stepped cone has a radius of 12 inches, and the radius of the corresponding driven pulley is 4 inches ; the distance between centres of pulleys is 3 feet, and there are three other pulleys on the driving-cone, having radii of 9, 6, and 3 inches respectively. It is required to find the radii of the corresponding pulleys on the driven cone. It will be necessary first to calculate the length of belt required, which is 10.334 feet, or about 10 feet 4 inches.

Next find whether the pulleys on the driving or driven cone are the largest. Half the length of belt is 5.167 feet. For the 9-inch pulley, 0.75 multiplied by 3.1416 is 2.356, and adding 3, the sum is 5.356, which is greater than 5.167, showing that the 9-inch pulley is larger than the pulley to be determined. For the 6-inch pulley : 0.5 multiplied by 3.1416 equals 1.571, and increased by 3 equals 4.571, and as this is less than 5.167, the 6-inch pulley is smaller than the pulley to be determined. Of course, then, the remaining 3-inch pulley is still smaller than the corresponding pulley in the driven cone.

To find the radius of the pulley corresponding to the one on the driving-cone whose radius is 9 inches : (1) 0.75 multiplied by 6.2832, 4.712. Subtracting 4.712 from 10.334, the remainder is 5.622 ; 5.622 divided by 3, 1.874. (2) 0.467 added to 1.874, 2.341. Square root of 2.341, 1.53. (3) Subtracting 1.53 from 1.571, the remainder is 0.041 ; 0.041 multiplied by 3, 0.123. (4) Subtracting 0.123 from 0.75, the remainder is 0.627 feet, or about 7½ inches, radius of required pulley.

Pulley corresponding to driving-pulley whose radius is 6 inches : 0.5 multiplied by 6.2832 equals 3.142. This subtracted from 10.334 equals 7.192, which divided by 3 equals 2.397—(1). Adding 0.467, we have 2.864, the square root of which is 1.692 —(2). 1.692 diminished by 1.571 equals 0.121, which multiplied by 3 equals 0.363—(3). Adding 0.5 gives 0.863 feet, or about 10$\frac{11}{32}$ inches, required radius.

Pulley corresponding to driving-pulley whose radius is 3 inches : 0.25 multiplied by 6.2832 equals 1.571. This, subtracted from 10.334, equals 8.763. Dividing the last by 3 gives 2.921— (1); and by adding 0.467 we have 3.388. Of this the square root is 1.841—(2). Subtracting 1.571 gives 0.27, which multiplied by 3 equals 0.81, and adding 0.25 gives 1.06 feet, or about 12$\frac{1}{16}$ inches, required radius.

The radii of the several pulleys on the two cones then will be :

Driving-cone..................12, 9, 6, 3 inches.
Driven cone.................... 4, 7$\frac{1}{2}$, 10$\frac{11}{32}$, 12$\frac{1}{16}$ inches.

B.

PULLEYS, Set-screws for.—These should be made of cast-steel with hollow points ; the ends should then be beveled to an edge surrounding the hole, and tempered to a dark straw. When set up, these screws cut circular indentations on the shaft, and exert an enormous force of resistance.

PULLEYS, Working value of.—Pulleys covered with leather, iron pulleys polished, and mahogany pulleys polished, rank for working value as 36, 24, and 25 per cent respectively, wood and iron uncovered being almost identical.

RAWHIDE BOXES for machinery.—A practical machinist says : "I have run a piece of machinery in rawhide boxes for fourteen years without oil ; it is good yet, and runs at 4500 per minute. I put it in while soft, and let it remain until dry."

SHAFTING ACCIDENTS, Preventing.—Accidents are common in large manufactories through the engagement of some portion of a workman's garments with a swiftly rotating shaft. A simple way of rendering these casualties impossible is to cover the shaft with a loose sleeve along its whole length. The sleeve may be of tin or zinc, and made so as to be removable if desired. The friction between it and the shaft would be sufficient to cause its rotation with the latter, but of course, in event of a fabric becoming wrapped around it, it would quickly stop, and allow of the easy extrication of the same. The sleeve should be lined with leather both within and at the ends in order to prevent noise. The same idea in the shape of loose covers might readily be applied to cog-wheels or pulleys.

SHAFTING, Lining.—Every one operating long lines of shafting should provide an adjusting-rod, as shown in the engraving. A may be a rod or a piece of gas-pipe, of sufficient length to reach from the shaft, O, to within about 4 feet of the floor ; an offset piece, B, is fixed to the top of this rod, which carries a right and left hand screw, C ; two jaws, D, travel upon this screw, one upon the right and the other upon the left hand thread, as shown. The screw may be worked by a ¼-in. wire, E, with a crank, F, at its lower end ; if a gas-pipe is used, the wire may pass through the pipe, and the lower end of the screw, C, enter the top of the pipe as a bearing. If the rod, A, is of wood, three or four wire staples will suffice as guides for the wire, as indicated. A target, G, with a clamp-screw, slides upon the rod, for the purpose of easy adjustment to the sights of the leveling instrument.

Now it will of course be apparent to every one that, whenever several sizes of shafting occur in the same line, this adjusting-rod will always give the exact central distance, O, of the shaft from the target ; hence we have only to plant the leveling instrument in a position to command a view of the target when suspended from each of the several bearings of a line of shafting, in order to adjust the level of a line with the utmost expedition and accuracy. An engineer's tripod and level is, of course, the best instrument for this purpose, but, when this is not at hand, an ordinary builder's level may be used : the longer it is, the better. Fix a temporary sight at one end of the level ; a piece of tin (with a small pin-hole) next the eye, and a piece of tin or thin wood with a large hole at the farther end, with a vertical and a horizontal thread stretched across the hole, with their point of intersection the same distance above the level as the hole in the eye-piece. The level may be used upon a level stand or table, some five feet from the floor.

To adjust a line of shafting laterally, an adjusting-rod must of course be used horizontally in connection with a strong line, stretched as taut as possible, at such distance from the shafting as to need nearly the full length of the rod to reach it. The reason for placing the line at such a distance from the shaft is to prevent the difference in level between the line and the shaft from materially impairing the truth of the result. If the line is very long, it will sag so much that a plumb-line suspended from the measuring point of the target or rod may be necessary for perfect accuracy.

The jaws, D, should be so formed that they may be applied to the inside of boxes. Pivot-boxes are

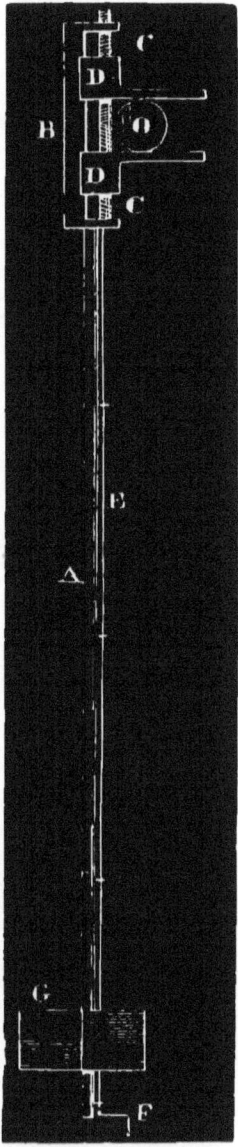

DEVICE FOR LINING
SHAFTING.

now so generally used, however, that this application of the rod is not so common.

SHAFTS, Sprung.—If a shaft springs in running, the trouble lies probably in either a too small diameter of the shaft for its weight and velocity, a set of unbalanced pulleys, or an unequal strain on either side by the belts.

WOOD, Lubricant for.—A mixture of black-lead and soap.

CEMENTS, GLUES, AND MOULDING COMPOSITIONS.

CASTS, To prepare plaster of Paris for.—Immerse the unburnt gypsum for 15 minutes in water containing 8 or 10 per cent of sulphuric acid, and then calcine it. Prepared in this way it sets slowly, but makes excellent casts, which are perfectly white instead of the usual grayish tint.

CASTS, Transparent.—Beautiful semi-transparent casts of fancy articles may be taken in a compound of 2 parts unbaked gypsum, 1 part bleached beeswax, and 1 part paraffine. This becomes plastic at 120°, and is quite tough.

CEMENTS.—*Air and water-tight, for casks and cisterns :* Melted glue 8 parts, linseed-oil 4 ; boil into a varnish with litharge. This hardens in 48 hours. *Plumbers' :* Black resin 1 part, brick-dust 2. Melt together. *For leaky boilers :* Powdered litharge 2 parts, fine sand 2, slaked lime 1. Mix with boiled linseed-oil. Apply quickly. *Acid-proof :* A paste of powdered glass and concentrated solution of water-glass. *Cutlers' :* (1) Pitch 4 parts, resin 4, tallow 2, and brick-dust 2. (2) Resin 4, beeswax 1, brick-dust 1. (3) Resin 16, hot whiting 1, wax 1. This is used for fastening blades in handles. *For ivory or mother-of-pearl :* Isinglass 1 part, white glue 2, dissolved in 30 parts hot water and evaporated to 6 parts. Add gum-mastic $\frac{1}{30}$ part, dissolved in $\frac{1}{2}$ part alcohol, and add 1 part zinc-white. Shake up and use warm. *Jeweler's, for uniting all substances :* Gum-mastic 5 or 6 bits as large as a pea dissolved in spirits of wine sufficient to render all liquid. In another vessel dissolve the same amount of isinglass in rum enough to make 2 ozs. of strong glue, adding 2 small pieces of gum ammoniacum, which must be moved until dissolved. Heat and mix the whole. Keep in a closely-corked phial, and put the latter in bo''ing water before using. *Black, for bottle-corks :* Pitch hardened by the addition of brick-dust and resin. *For jet :* Use shellac, warming the edges before applying, and smoke the joint to make it black. *For meerschaum or china :* (1) Make a dough of garlic, rub on the edges and bind tightly together. Boil the object for half an hour in milk. (2) Use quicklime mixed to a thick cream with white of egg. *Soft, for steam-boilers :* Red or white lead in oil 4 parts, iron borings 3 parts. *Gasfitters' :* Resin 4½ parts, wax 1, Venetian red 3. *Coppersmiths' :* Boiled linseed-oil and red lead made into a putty. This is used to secure joints and on washers. *For emery on to wood :* Equal parts of shellac, white resin, and carbolic acid in crystals. Add the acid after the others are melted. *Iron and emery :* Coat the meta. with oil and white-lead, and when hard apply the emery mixed with glue. *French putty,* hard and permanent : Linseed-oil 7 parts, brown umber 4, boiled for 2 hours, $\frac{1}{4}$ part white-wax stirred in. Remove from fire and thoroughly mix in white-lead 11, and fine chalk 5$\frac{1}{2}$ parts. *India-rubber :* Fill a bottle $\frac{1}{10}$ full of native india-rubber cut into fine shreds. Pour in benzole from coal-tar till the bottle is $\frac{3}{4}$ full. The rubber will swell ; and if the whole be shaken every few days, the mixture will become as thick as honey. If too thick, add benzole ; if thin, add rubber. This dries in a few minutes, and will unite backs of books, straps, etc., very firmly. *Chinese, for fancy articles, wood, glass, etc. :* Finest pale-orange shellac, broken small, 4 parts, rectified spirit 3 parts. Keep in a corked bottle in a warm place until dissolved. It should be as thick as molasses. *Rust joints :* (1) Clean iron borings 2 parts, flowers of sulphur $\frac{1}{16}$, sal-ammoniac $\frac{1}{16}$. (2) Finely-powdered iron borings 1 part, sal-ammoniac $\frac{1}{8}$, flowers of sulphur $\frac{1}{16}$. Pound together and keep dry. For use, mix 1 part with 20 of pounded iron borings, and mix to a mortar consistence with water. *For making metallic joints sound :* (1) Use a putty of boiled linseed-oil and red-lead. (2) Use a putty of equal parts of white and red lead. *For electrical and chemical apparatus :* Resin 5 parts, wax 1, red ochre 1, plaster of Paris $\frac{1}{8}$. Melt at moderate heat. *For mending stone, or as mastic for brick walls :* Make a paste of linseed-oil with clean river sand 20 parts, litharge 2, quicklime 1. *For chucking work in the lathe :* (1) Black resin 8 parts, yellow wax 1 ; melt together. For use, cover the chuck to $\frac{1}{16}$ in. thick, spreading over the surface in small pieces, mixing it with $\frac{1}{4}$ its bulk of gutta-percha in thin slices. Heat an iron to dull red and hold it over the chuck till the mixture and gutta-percha are melted and liquid. Stir the cement with the iron until it is smoothly mixed. Chuck the work, lay on a weight to enforce contact, and let it rest for half an hour before using. (2) Burgundy pitch 2 parts, resin 2, yellow wax $\frac{1}{8}$, dried wax 2. Melt and mix. (3) Resin 4 parts, melted with pitch 1. While boiling add brick-dust until dropping a little on stone shows the mixture to be sufficiently hard. *Elastic,* for leather or india-rubber : Bisulphide of carbon 4 ozs., shredded india-rubber 1 oz., isinglass 2 drachms, gutta-percha $\frac{1}{4}$ oz. Dissolve, coat the parts, dry, then heat the layer to melting, place and press the parts together. *Water-tight, for wooden vessels :* Lime, clay, and oxide of iron, mixed, kept in a close vessel and compounded with water for use. *For leather, straps, etc. :* Gutta-percha dissolved in bisulphide of carbon. Keep tightly corked and cool. It should be of the consistence of molasses. *For marble, or for attaching glass to metal :* Plaster of Paris soaked in a saturated solution of alum and baked hard. Grind to powder and mix with water for use. Can be colored to imitate any marble, and takes a fine polish. *Impervious, for corks, etc. :* Zinc-white rubbed up with copal varnish. Give two coats so as to fill all the pores, and finish with varnish alone. *For cracks in wood :* (1) Slaked lime 1 part, rye meal 2, and linseed-oil 2. (2) Use a paste of sawdust and prepared chalk with glue 1 part, dissolved in water 16. (3) Oil-varnish thickened with equal parts of litharge, chalk, and white and red lead. *For wood and glass or metals :* (1) Resin and calcined plaster, the former melted, made into a paste. Add boiled oil to consistence of honey. (2) Dissolved glue and wood-ashes to consistence of varnish. *Fireproof and water-proof :* Pulverized zinc-white, sifted peroxide of manganese, equal parts. Make into a paste with soluble glass. *To mend iron pots and pans :* Partially melt 2 parts sulphur, and add 1 part fine black-lead. Mix well, pour on stone, cool, and break in pieces. Use like solder with an iron. *London cement, for glass, wood, china, etc.:* Boil a piece of cheese three times in water, each time allowing the water to evaporate. Mix the paste left with quicklime. *For aquaria :* (1) For fresh water aquaria : Take ½ gill gold-size, 2 gills red-lead, 1½ gills litharge, and sufficient silver sand for a thick paste. This sets in about 2 days. (2). For fresh or salt water : Take ½ gill powdered resin, 1 gill dry white sand, 1 gill litharge, 1 gill plaster of Paris. Sift ; and for use mix with boiled linseed-oil to which a little dryer has been added. Mix 15 hours before using, and allow 2 or 3 hours to dry. *For petroleum lamps,* impervious to the oil : Resin 3 parts, boiled with water 5 and caustic soda 1. Then mix with half its weight of plaster of Paris. This sets in ¾ hour. *Roman :* Green copperas 3½ lbs., slaked lime 1 bushel, fine gravel sand 1 bushel. Dissolve the copperas in hot water, and mix all to proper consistence. Keep stirred. *Glass to glass,* for sign-letters, etc. : Melt in a water-bath liquefied glue 5 parts, copal varnish 15, drying-oil 5, oil of turpentine 2, turpentine 3. Add slaked lime 10. *Hydraulic :* Oxide of iron 1 part, powdered clay 3, and boiled oil to a stiff paste. *Stone :* Sand 20 parts, litharge 2, quicklime 1, mixed with linseed-oil. *Leather and cloth,* for uniting parts of boots and shoes, seams, etc. : Gutta-percha 16 parts, india-rubber 4, pitch 2, shellac 1, oil 2. Mix and use hot. *Mahogany :* Shellac melted and colored. *Colorless, for paper :* Add cold water to rice-flour, mix, bring to proper consistence with boiling water, and boil one minute. *Water-proof, for cistern stones :* (1) Whiting 100 parts, resin 68, sulphur 18½, tar 9. Melt together. (2) Sand 100 parts, quicklime 28, bone ashes 14, mixed with water. *Transparent :* India-rubber 75 parts, chloroform 60. Mix, and add mastic 15. *Cloth to iron :* Soak the cloth in a dilute solution of galls, squeezing out the superfluous moisture, and applying the cloth, still damp, to the surface of the iron, which has been previously heated and coated with strong glue. The cloth should be kept firmly pressed upon the iron until the glue has dried. *For cracks in stoves :* Finely-pulverized iron (procured at a druggist's) made into a thick paste with water-glass. The hotter the fire, the more the cement melts and combines, and the more completely does the crack become closed. *For china, glass, etc. :* Diamond cement, for glass or china, is nothing more than isinglass boiled in water to the consistence of cream, with a small portion of rectified spirit added. It must be warmed when used. 2. White-lead rubbed up with oil. Articles mended with this must stand for a month. *For corks of benzine-bottles :* A paste of concentrated glycerine (commonest kind) and litharge. This soon hardens, and is insoluble in benzine or any of the light hydro-carbon oils. *For caustic lye tanks:* The tanks may be formed of plates of heavy-spar, the joints being cemented together by a mixture of 1 part finely divided india-rubber dissolved in 2 parts turpentine oil, with 4 parts powdered heavy-spar added. *Colored :* Soluble glass of 33° B. is to be thoroughly stirred and mixed with fine chalk and the coloring matter well incorporated. In the course of six or eight hours a hard cement will set. The following are the coloring materials : 1. Black : Well-sifted sulphide of antimony. This can be polished with agate to a metallic lustre. 2. Gray-black : Fine iron-dust. 3. Gray : Zinc-dust. This has a brilliant lustre, and may be used for mending zinc castings. 4. Bright green : Carbonate of copper. 5. Dark green : Sesquioxide of chromium. 6. Blue : Thénard's blue. 7. Yellow : Cadmium. 8. Bright red : Cinnabar. 9. Violet red : Carmine. 10. Pure white : Fine chalk as above.

CEMENT, PORTLAND, To test.—Three tests are used : (1) Resistance to tensile force. (2) Specific gravity. (3) Water test. The first is by making a specimen briquette in a mould with a transverse section of 2.25 square inches, the specimen being held vertically in clips, which is placed under the short arm of a steel-balance, and broken. A test of 500 lbs. has been used on an area of 2.25 square inches after 7 days' immersion in water. The second method is by finding the weight in pounds of the struck bushel. The water test is useful when the others can not be applied. It consists of gauging a small quantity of the dry powder with water, and immediately immersing it in water. If the sharper edges crack or break away after a short time, the cement is too hot or fresh, or is inferior in quality. The weight of good Portland cement ranges from 100 lbs. to 130 lbs. per bushel, equal to from 80 lbs. to 102 lbs. per cubic foot. The lighter kinds set more rapidly than the heavier, but are weaker. The specific gravity should be of 110 lbs. to a bushel.

GLUED JOINTS, Strength of.—The absolute strength of a well-glued joint is given as follows in pounds per square inch :

	Across the grain.	With the grain.
Beech,	2133	1095
Elm,	1436	1124
Oak,	1735	568
Whitewood,	1493	841
Maple,	1422	896

It is customary to use from $\frac{1}{6}$ to $\frac{1}{10}$ of the above values to calculate the resistances which surfaces joined with glue can permanently be submitted to with safety.

GLUE, Fire-proof.—A handful of quicklime mixed in 4 ozs. of linseed-oil and boiled to a good thickness makes, when spread on plates and hardened, a glue which can be used in the ordinary way, but which will resist fire.

GLUE, Liquid.—Dissolve the glue in an equal amount of strong hot vinegar, adding $\frac{1}{4}$ alcohol and a little alum. Will keep indefinitely.

GLUES, Marine.—(1) Pure india-rubber 1 pt., dissolved by heat in naphtha ; when melted, add shellac, 2 pts. (2) Glue, 12 pts., water to dissolve, and yellow resin, 3 pts. Melt, add turpentine, 4 pts., and mix. *Portable, for draughtsmen :* Glue 5 ozs., sugar 2 ozs., water 8 ozs. Melt in water-bath, cast in moulds, and dissolve for use in warm water. *For bank-notes :* Fine glue or gelatine, 1 lb., dissolved in water, and the water evaporated until nearly expelled. Add ¼ lb. brown sugar, and pour in moulds. *For gutta-percha :* Common black pitch 2 pts., gutta-percha 1 pt. Mould into shape. *Elastic :* Dissolve glue in a water-bath, evaporate to a thick fluid, and add an equal weight of glycerine. Cool on a slab. *Liquid :* (1) White glue 16 ozs., dry white-lead 4 ozs., soft water 2 pints, alcohol 4 ozs. Stir and bottle while hot. (2) Glue 3 pts. softened in 8 parts water. Add ½ pt. muriatic acid and ¾ pt. sulphate of zinc. Heat to 176° Fahr. for 12 hours. Allow the compound to settle. *Heat and moisture-proof :* Linseed-oil 4 ozs., 1 handful of quicklime ; boil to good thickness, and cool. It will become very hard, but is as easily dissolved as common glue. *Water-proof, simple :* Common glue 1 lb. boiled in 2 qts. skimmed milk.

GLUE, Test for goodness of.—Assuming that that is the best glue which will take up most water, take 50 grains of the specimen and dissolve it in 3 ozs. water in a water-bath. When dissolved, set it by for 12 hours to gelatinize, and then take an ounce chip-box, place it on the surface of the gelatine, and put shot into the box until it sinks down to a mark on the outside. It will be found that the stronger the glue, the more shot it will take to sink the box down so that the mark shall be level with the surface of the gelatine. In a trial with very fine glue, 50 grains of glue dissolved and gelatinized with 3 ozs. of water, supported to the mark on the box 6 ozs. of shot, at a temperature of 58° Fahr. On trying the same experiment with best Russian isinglass, 9 ozs. of shot were supported, the temperature being the same. This test is of course intended as a comparative one between two kinds of glue, or between any kind taken as a standard and another compared with it. The placing of the mark is arbitrary.

GLUE, To bleach.—Soak in moderately strong acetic acid for two days, drain, place on a sieve, and wash well with cold water. Dry on a warm plate.

MOULDING ARCHITECTURAL ORNAMENTS.—A good composition for this purpose is made of chalk, glue, and paper-paste.

MOULDING COMPOSITION.—Five parts of sifted whiting mixed with a solution of 1 part glue, together with a little Venice turpentine to obviate the brittleness, makes a good plastic material, which may be kneaded into figures or any desired shape. It should be kept warm while being worked. It becomes as hard as stone when dry.

PASTE, To mould figures in.—Take the crumb of a new-drawn white loaf, mould in a mass until the whole becomes as close as wax and very pliable. Then heat and roll with a rolling-pin. Mould it to the required shape, and dry in a stove.

PHOTOGRAPHIC PRINTS, To varnish.—Heat a piece of glass, and rub a little wax over it with a bit of cotton-wool. Pour water over the plate, and press the picture down upon it with a piece of filtering-paper. When dry, the picture is removed, and will be found to possess a brilliant surface.

PICTURE-FRAMES, Composition for.—(1.) To make composition ornaments for picture-frames : Boil 7 lbs. best glue in 3½ pints water, melt 8 lbs. white resin in 3 pints raw linseed-oil ; when the ingredients are well boiled, put them into a large ves- sel and simmer them for half an hour, stirring the mixture and taking care that it does not boil over. When this is done, pour the mixture into a large quantity of whiting, previously rolled and sifted very fine, mix it to the consistence of dough, and it is ready for use. (2) Dissolve 1 lb. glue in 1 gallon water ; in another kettle boil together 2 lbs. resin, 1 gill Venice turpentine, and 1 pint linseed-oil ; mix together in one kettle, and continue to boil and stir them together till the water has evaporated from the other in- gredients ; then add finely-pulverized whiting till the mass is brought to the consistence of soft putty. This composition will be hard when cold ; but being warmed, it may be moulded to any shape by carved stamps or prints, and the moulded figures will soon become dry and hard, and will retain their shape and form permanently. Frames of either material are well suited for gild- ing.

PLASTER CASTS, To toughen.—Immerse in a hot solution of glue long enough for the mass to be well saturated. They will bear a nail driven in without cracking.

PLASTER MODELS, Mending.—Sandarac varnish is the best ma- terial. Saturate the broken surfaces thoroughly, press them well together, and allow them to dry.

PLASTER MOULDS.—Glycerine is said to be a good coating for the interior, but practical plaster moulders still use, as of old, a mixture of lard and oil.

METAL-WORKING HINTS AND RECIPES.

ALLOY for filling defects in small castings.—Lead 9 parts, antimony 2, bismuth 1. This expands on cooling.

ALLOY OF COPPER, which will attach itself to glass, metal, or porcelain.—20 to 30 parts finely blended copper (made by reduc- tion of oxide of copper with hydrogen or precipitation from solu- tion of its sulphate with zinc) are made into a paste with oil of vitriol. To this add 70 parts mercury and triturate well ; then wash out the acid with boiling water and allow the compound to cool. In 10 or 12 hours, it becomes sufficiently hard to receive a brilliant polish and to scratch the surface of tin or gold. When heated it becomes plastic, but does not contract on cooling.

ALLOY, " OROIDE."—This is made of pure copper 100 parts, tin 17 parts, magnesia 6 parts, sal-ammoniac 3½ parts, quicklime 1½ parts, tartar of commerce 9 parts. The copper is first melted, then the magnesia, sal-ammoniac, lime, and tartar in powder are added little by little and briskly stirred for half an hour. The tin is lastly mixed in grains until all is fused. The crucible is covered, and the fusion maintained for 35 minutes, when the dross is skimmed off and the alloy is ready for use.

ALLOYS, To extract silver from old.—Dissolve in nitric acid and precipitate the chloride of silver with a solution of common salt. The silver is reduced to a pure state by mixing the chloride with an equal weight of bicarbonate of soda and smelting in a common sand crucible.

ALUMINUM SILVER.—The following alloy is distinguished by its beautiful color, and takes a high polish : Copper 70, nickel 23, aluminum 7, total 100.

BABBITT METAL.—There are a large number of recipes for this alloy, but the following gives an excellent composition for general use : Tin 50 parts, antimony 5 parts, copper 1 part.

BELL, CRACKED, To repair.—A cracked bell which gives a jarring sound may be improved by sawing or filing the ruptured edges so that they are not brought together by the vibration of the blow.

BOILER-TUBES, IRON, To preserve.—A coating of red-lead and boiled linseed-oil, applied to iron boiler-tubes, acts as a great preservative.

BRASS, Black stain for.—Arsenious acid 2 parts, hydrochloric acid 4, sulphuric acid 1, water 80.

BRASS SCRAP, To utilize.—The best way is to melt it in with new brass, putting it in with the zinc after the copper is melted.

BRASS, To blacken.—Mix 4 parts hydrochloric acid and 1 part arsenic (by weight) ; put on bright, dry, and lacquer.

BRASS, To clean.—Rub bichromate of potash fine, pour over it about twice the bulk of sulphuric acid, and mix this with an equal quantity of water. The dirtiest brass is cleaned by this in a trice. Wash the metal immediately after in plenty water ; wipe, rub dry, and polish with powdered rottenstone.

BRASS, VERT DE BRONZE ON, To produce.—Dissolve 2 ozs. nitrate of iron and 2 ozs. hyposulphite of soda in 1 pint water. Immerse the articles till they are of the required tint, as almost any shade from brown to red can be obtained ; then wash well with water, dry, and brush. One part perchloride of iron and 2 parts water mixed together, and the brass immersed in the liquid, gives a pale or deep olive-green, according to the time of immersion. If nitric acid is saturated with copper, and the brass dipped in the liquid and then heated, the article assumes a dark-green color.

BRONZE for gongs and cymbals.—This is made with 20 per cent of tin, and is hammered into shape while at a red heat ; it is then tough and malleable, but is very brittle when cold.

BRONZE for small castings.—Fuse together 95 parts of copper and 36 parts of tin.

BRONZE, GREEN.—The bluish-green bronze used for ornamental articles is made of any metal, first covered with a varnish made of ground tin or bronze powder rubbed up with honey in gum-water. Then wash with a mixture composed of sal-ammoniac ½ oz., common salt ¼ oz., and 1 oz. spirit of hartshorn in 1 pint vinegar. After applying the mixture, leave for a day or two in the sun, and then, if necessary, add a second coat. This is a good way to renovate old gas-fixtures.

BRONZE, JAPANESE.—A curious bronze is produced in Japan, which, when made in thin plates, resembles slate, and is covered with designs in silver. It contains, in addition to copper, from 4 to 5 per cent of tin, and on an average 10 per cent of lead. The combination is easily moulded into thin plates. These are varnished, and through the covering the designs are scratched with a burin. The plate is then plunged in a silver-bath, when the silver is deposited on the unprotected portions. Lastly, it is placed in a muffle-furnace, when the copper blackens and the silver remains bright.

BRONZING HARDWARE.—Brown bronze dip, for coating hat-hooks and similar small hardware articles, is made of iron scales 1 lb., arsenic 1 oz., muriatic acid 1 lb., zinc, solid, 10 ozs. The zinc should be kept in only when the bath is used. The castings must be perfectly free from sand and grease.

CASE-HARDENING, to be quickly performed, is done by the use of prussiate of potash. This is powdered and spread upon the surface of the iron to be hardened, after the iron is heated to a bright red. It almost instantly fluxes and flows over the surface ; and when the iron is cooled to a dull red, it is plunged in cold water. Some prefer a mixture of prussiate of potash 3 parts, sal-ammoniac 1 part ; or prussiate 1 part, sal-ammoniac 2 parts, and finely-powdered bone-dust (unburned) 2 parts. The application is the same in each case. Proper case-hardening, when a deep coating of steel is desired, is done by packing the article in an iron box with horn, hoof, bone-dust, shreds of leather or raw hide, or either of these, and heating to a red heat for from 1 to 3 hours, then plunging the box into water.

CHAIN, Strength of.—To ascertain the strength of short-linked chains : (1) Multiply the square of the diameter (reckoned in six-teenths of an inch) by .035 ; the product will be the weight the chain will support in tons. (2) The square of the diameter in eighths of an inch = weight of chain in lbs. per fathom. The square of the diameter in eighths ÷ 2 = breaking weight in tons. Thus for a chain made of $\frac{3}{8}$ iron, the weight = $3^2 = 9$ lbs. per fathom, and its breaking weight would be $\frac{3^2}{2} = 4\frac{1}{2}$ tons. The utmost load put upon it should not exceed $1\frac{1}{2}$ tons, the safe constant load being 18 to 20 cwt.

COLORING METALS.—Take hyposulphite of soda 4 ozs., dissolved in $1\frac{1}{2}$ pints of water ; add a solution of 1 oz. acetate of lead in same quantity of water. Articles to be colored are placed in the mixture, which is then gradually heated to boiling. The effect of the solution is to make iron resemble blue steel ; zinc becomes bronze, and copper or brass becomes successively yellowish-red, scarlet, deep blue, bluish-white, and finally white with a tinge of rose. The solution has no effect on lead or tin. By replacing the acetate of lead in the solution with sulphate of copper, brass becomes of a fine rosy tint, then green, and finally of an iridescent brown color. Zinc does not cover in this solution; but if boiled in a solution containing both lead and copper, it becomes covered with a black crust, which may be improved by a thin coating of wax.

COLUMNS, Strength of hollow.—The hollow cylinder is the strongest form of section under compressive force. The experiments by which this was proved were conducted upon hollow tapering columns of cast-iron, upon cross-sections, as used in the connecting-rods of steam-engines, and upon forms in which the metal was cast in the shape of the letter H. All these forms proved considerably weaker than the hollow cylinder of equal weight of metal. As the relative merits of these forms of casting metal are of constant use, we append their proportionate strengths: Hollow cylindrical pillar, 100; H-shaped pillar, 75; +-shaped pillar, 44. The examples were all of the same weight and length, with rounded ends. General Morin's rule for the thickness of cast-iron pillars may be relied upon, as it is based upon the founder's experience of the minimum thickness. Height, feet, 7 to 10, 10 to 13, 13 to 20, 20 to 27.; minimum thickness, inch, 0.5   0.6   0.8   1.0

Another rule is to make the thickness in no case less than $\frac{1}{12}$ of the diameter. Cellular or tubular girders exemplify to a still greater degree the value of hollow construction.

COPPER AND BRASS, Coating, with zinc.—Dip the articles into a boiling concentrated solution of sal-ammoniac containing finely-divided zinc.

COPPER-WELDING.—A good welding mixture is composed of phosphate of soda 358 parts, boracic acid 124 parts.

CRUCIBLES.—The best crucibles are composed of the following compositions, which are of two kinds—namely, with and without plumbago. 3 parts by measure of the Stourbridge best crucible clay, 2 parts cement, consisting of old used-up fire-bricks, and 1 part hard coke. These ingredients must be ground and sifted through a $\frac{1}{8}$ in. mesh sieve; the sieve must not be finer, otherwise the pot will crack. This composition must be mixed with sufficient clean cold water, trodden with the bare foot to the consistency of stiff dough and allowed to stand for 3 or 4 days, well covered with damp cloths, to admit of its sweating and the particles of clay becoming thoroughly matured. It is then ready for use, and must be blocked by hand on a machine. Owing to the coarseness of this composition, the pot can not well be thrown on the potter's wheel; and in no instance can it be made by pressing. The crucible must not be burnt in a kiln, but merely highly and thoroughly dried before being placed in the furnace for use. For brass and copper melting, it will stand one good hard day's work; but care must be taken to replace the pot again in the furnace after the metal has been poured. If the pot be not allowed to go cold, it will last for several days. It will, with the greatest safety, stand one melting of wrought-iron. The cost, when made on the steel manufacturer's own premises, is about forty cents per pot, each pot holding from 100 to 120 pounds of metal. Good Hessian crucibles are composed of 2 parts of the best German crucible clay and 5 parts pure fine quartz sand. This composition must be sifted through a $\frac{1}{4}$ in. mesh sieve ; it is then tempered and trodden with the bare foot, as before described. When ready for use, it is pressed into different sizes of crucibles, which, when thoroughly dry, are placed in the kiln or furnace and burnt hard.

Another composition : 2 parts best Stourbridge crucible clay, 3 parts cement ; sift through a $\frac{1}{4}$-in. sieve ; temper as before described and block by hand on the machine. When thoroughly dry, it is placed in the kiln and burnt hard. These crucibles are principally used for melting gold and silver, and also for dry analysis. The best and most perfect fire-clay for crucible making is nearly always found in the pavement of coal. Some of the Pittsburg fire-clays, and those found to exist in the pavements of some of the Pennsylvania coal-mines, are excellent fire-clays. But the various compositions can not be described, as they are as numerous as the different kinds of clay. The Birmingham soft tough pot consists of 2 parts of the best Stourbridge crucible clay, 3 parts plumbago, and 1 part cement, consisting of old used-up crucibles ground and sifted through a $\frac{1}{4}$ in. mesh sieve.

Another composition : 4 parts of the best Stourbridge crucible clay, 3 parts plumbago, 2 parts hard coke, and 1 part cement, consisting of old pots ground and sifted as before. Where old pots can not be had, the above composition must be burnt hard, ground, and sifted. The scales or chippings of the insides of gas-retorts are far superior to the best common hard coke. But where scales and chippings can not be had, hard coke is the best substitute. All the ingredients of this composition must be sifted through a $\frac{1}{8}$ in. sieve (but not finer), tempered, and made as before described. When thoroughly dry, it is placed in the kiln and annealed, but not burnt hard. This composition makes a pot (for melting the hardest metal) which can not be melted at any pitch of heat, nor can it be cracked with the most sudden heating and cooling. It is regularly known to stand 14 and 16 meltings of iron—even wrought-iron. Any steel manufacturer can make the pot on his own premises at a cost of $1.20 or thereabouts, the pot holding from 100 to 120 lbs. of metal.

ETCHING UPON STEEL.—Warm the steel, and rub on a coating of white-wax or hard tallow. When hard, mark the device through the wax with a sharp-pointed tool ; apply nitric acid, and allow it to stand for a few minutes ; then wash off the acid thoroughly with water, heat the steel, and rub off the wax with a rag. The device will be found etched on the steel.

GOLD AND SILVER, Test for.—A good test for gold or silver is a piece of lunar caustic, fixed with a pointed stick of wood. Slightly wet the metal to be tested, and rub it gently with the caustic. If gold or silver, the mark will be faint ; but if an inferior metal, it will be quite black.

GUN-BARRELS, To bronze.—Clean thoroughly, and apply (by means of a rag) nitric or sulphuric acid diluted with its volume of water.

HARDENING PICKLE.—Spring-water made into a brine strong enough to float an egg, then boiled to precipitate the lime, and allowed to cool.

IRON ARTICLES, Brightening.—When taken from the forge or rolls, the articles are placed in dilute sulphuric acid (1 to 20) for an hour ; they are then washed clean in water, dried with saw-dust, dipped for a second or so in nitrous acid, washed and dried as before, and finally rubbed clean.

IRON RINGS, Welding, without scaling.—Take iron filings 1000 parts, borax 500 parts, resinous oil of any kind 50 parts, sal-ammoniac 75 parts. Pulverize completely and mix ; heat the rings to a cherry red, powder the parts with the mixture, and join them together.

IRON, Simple fire-plating for.—By rubbing the surface of iron or other metals with soda amalgam, and then pouring over it a concentrated solution of chloride of gold, the gold is taken up by the amalgamated surface, and it is only necessary to drive off the mercury with the heat of a large lamp to obtain a fine gilded sur-face that will bear polishing. By writing or drawing a design on the iron, the drawing will be re-produced in pure gold. Silver and platinum salts are said to act in a similar manner to the gold.

IRON, To gild cast.—The cheapest way is to use bronze or mosaic gold. The castings are first to be heated hotter than the hand can bear, but not so hot as to burn the varnish, and coated with mosaic gold mixed with a small quantity of alcohol varnish. If the iron is polished, it must be heated previously and rubbed over with a rag dipped in vinegar.

LEAD, Determining presence of, in tin vessels.—The metal to be tested is first touched with nitric acid and then heated, when the acid evaporates. If lead be contained, stannic acid and nitrate of lead remain. Iodide of potassium is then applied, forming yel-low iodide of lead ; while the stannic acid is white. The yellow stain, therefore, indicates lead, the white, tin.

JEWELRY, To restore the lustre of.—Take 1 oz. cyanide potas-sium and dissolve in 3 gills water. Attach the article to be cleansed to a wire hook, immerse and shake in the solution for a second or two, and remove and wash in clean water, then in warm water and soap. Rinse again, dip in spirits of wine, and dry in boxwood sawdust. If the solution is kept, put it in a tightly-corked bottle, and label POISON conspicuously. One caution is necessary : Do not bend over the solution so as to inhale the odor, nor dip the fingers in it ; if one of the articles drops from the hook, better empty the solution into another vessel.

METAL SURFACES, To protect, from moisture.—Inclose them in tight compartments containing lumps of quicklime.

MINERALS AND METALS, Hardness and tenacity of.—In mine-ralogy, in which science the hardness is an important characte-ristic, ten bodies are usually taken as points of comparison—the softest being termed 1 and the hardest 10. These are: 1, talc ; 2, gypsum ; 3, carbonate of lime ; 4, fluor-spar ; 5, phosphate of lime ; 6, felspar ; 7, quartz ; 8, topaz ; 9, corundum ; 10, diamond. Hence, when scientific works speak of the hardness of a body being 6, 8, 4, etc., reference is made to the relative hardness expressed by the list above given.

The tenacity of metals is estimated by the resistance which wires of the same diameter experience when passed at equal temperature through the same hole of a draw-bench. The following table gives the relative tenacity of various metals and alloys : Steel already drawn, 100 ; iron already drawn, 88 ; brass already drawn, 77 ; gold at 0.875, annealed, 73 ; steel annealed, 65 ; copper already drawn, 68 ; silver at 0.750, annealed, 58 ; silver at 0.875, 54 ; brass annealed, 46 ; iron annealed, 42 ; platinum annealed, 38 ; copper annealed, 38 ; fine gold annealed, 37 ; fine silver annealed, 37 ; zinc, 34 ; tin, 11 ; lead, 4.

PLATINUM-BRONZE.—This is made of nickel 100 parts, tin 10, platinum 1. It is entirely unoxidizable, and especially adapted for cooking-utensils.

QUICKSILVER, Coating iron with.—Clean the iron first with hydrochloric acid, then immerse it in a dilute solution of sulphate of copper mixed with a little hydrochloric acid, by means of which it becomes covered with a slightly-adherent layer of copper. It is then to be brought into a very diluted solution of mercurial sublimate mixed with a few drops of hydrochloric acid. The article will become covered with a layer of mercury, which can not be removed even by rubbing. This is good as a protection from rust.

SADIRONS, Finishing.—See that your buff-wheels are well-balanced after they are covered. Let the wheel be covered with thick leather before covering with emery. Get as good a surface on the article as you can from a wheel covered with No. 70 emery. Mix flour of emery with melted beeswax, and stir in till it is thick. When the mass is cool, rub it on a newly-covered wheel with No. 80 emery. Then set the wheel running, and hold on a flint to smooth it until the surface is sufficiently fine to suit.

SILVER ORNAMENTS, Imitation.—Ordinary plaster models are covered with a thin coat of mica powder, which perfectly replaces the ordinary metallic substances. The mica plates are first cleaned and bleached by fire, boiled in hydrochloric acid, and washed and dried. The material is then finely powdered, sifted, and mingled with collodion, which serves as a vehicle for applying the compound with a paint-brush. The objects thus prepared can be washed in water, and are not liable to be injured by sulphuretted acids or dust. The collodion adheres perfectly to glass, porcelain, wood, metal, or *papier maché*.

SILVER, Producing satin finish on, by sand-blast.—The following is the method adopted at a large silver-plating establishment : Air is compressed by the driving-engine of the works into an ordinary reservoir, and thence distributed through pipes which extend along the front of the workmen's tables ; and above the latter is a sand receptacle, V-shaped, from which a stream of sand falls, and is met by a downward blast from the pipe, which current drives the material in a stream through a small hole in the table, beneath which a receptacle to receive the sand is placed. The workman, whose fingers are covered with rubber to protect them, holds the article in the jet and under the table, watching it through a pane of glass let into the top of the latter. The operation is necessarily very rapid, as the article has only to be turned so that the blast strikes the required portions, when the work is completed. The exposure to the jet, even for an instant, would cut through the Britannia, upon which the plating is afterward deposited. By the interposition of rubber screens of suitable shape, against which the sand has no abrading effect, any fancy patterns or letters are easily imprinted on the surface, the latter, of course, being satin-finished, while the spaces protected by the screens are afterward burnished.

SILVER, Restoring color to.—This is adapted to treating silver filagree ornaments, rendering them dead white. The process has long been a trade secret. If any pewter is found in the articles, it should not be attempted. Pound together charcoal 3 parts, and of nitre 1; add sufficient water to form a paste. With a camel's-hair brush give the article a thin coat of the mixture, put it in a small annealing-pan, and submit it to the fire until it becomes red-hot; then withdraw it from the fire, let it stand a minute, and turn it out into a weak solution of sulphuric acid (1 part acid, 10 parts water) in the boiling-pan. Boil, pour off the acid, rinse; wash with warm water and soap, using a soft brush; dip in spirits of wine, and dry in boxwood sawdust. If any spots should still remain on the work, anneal it without the mixture, boil out and wash as before. Burnish the parts intended to be bright. Do not use the common American saltpetre. The English nitre, although it costs more, is really less expensive, as a smaller quantity goes further and does the work more effectually. Purchase at the wholesale druggists.

SILVER, To regain, from broken black-lead crucibles.—Pulverize the crucible and digest it in nitric acid for several hours. Decant off the clear liquid, and add to it muriatic acid until no further precipitate forms. Allow to settle, and again decant the clear liquid, wash the precipitate several times with clean water, dry, and fuse in a small crucible with a quantity of carbonate of soda.

SLAG, Utilization of. Prussian method.—The high furnaces are provided with a continual overflow for the slag, which runs through a narrow gutter formed in the sand into a shallow pit, through which a small stream of water is kept running. By this chilling process the slag assumes the form of a fine gravel. An endless chain at once lifts the slag out of the pit and loads it upon cars. By grinding this material fine in a cement-mill, it is formed into an excellent sharp building-sand; the great bulk of it, however, is used, without further reducing its grain, for making bricks.

For this purpose it is mixed with one half of its bulk of mortar in a trough in which three shafts provided with long blades are revolving. It is then shoveled into the brick-machines, each of which turns out about twenty-five bricks a minute. These bricks are piled up in the open air for drying, and are ready for use after about six weeks. They continue to harden on exposure to the air, and are said to possess greater strength than ordinary burnt bricks. They are extensively used for all kinds of buildings, their light-gray color producing a very pleasing effect, and the roughness of their surface fitting them particularly well for retaining a coating of mortar. They can not be used, however, for foundation walls, as by the absorption of moisture their cohesiveness is impaired.

The most interesting process is the following : As a thin stream of the fluid slag, falling from a narrow gutter, passes the nozzle of the steam-pipe, a jet of steam is blown through it, and by this simple process it is solidified in the form of most delicate fibres, resembling asbestos or spun glass ; and it falls to the ground like a loose mass of grayish wool. This material is an excellent non-conductor of heat, and is used for covering steam-pipes, boilers, etc. The sole expenditure in its manufacture is that of the steam, the exact amount of which could not be ascertained. The material is sold for about $5 per cwt. The steam-pipe is about 1½ inches in diameter, and the nozzle is simply a pipe, flattened and then curved into a semicircular form, in order to give the most advantageous shape to the steam-jet. The steam used has a pressure of about 50 lbs. per square inch.

SOLDERING LIQUID.—Into hydrochloric acid place as much scrap-zinc as it will dissolve, still leaving a sponge of zinc. Use the mixture for soldering brass-work. To solder cast or wrought iron, add sal-ammoniac ; and for sheet-tin work, omit the sal-ammoniac.

SOLDER, Jewelers'.—Melt 1 part lead, add 2 parts tin, and throw in a small bit of resin as a flux. This is strong, easily flowing, and white. In soldering fine work, wet the parts to be joined with muriatic acid in which as much zinc has been dissolved as the acid will take up. It is cleaner than the old method of using Venice turpentine or resin.

SOLDER, Silver.—Put into a clean crucible, silver 2 parts, clean brass 1 part, with a small piece of borax. Melt and pour into ingots. Solder made from coin, as it frequently is, often meets with difficulty around the joints, requiring the use of the file to remove it, while the addition of any of the inferior metals to the solder causes it to eat into the article joined by it.

STEEL AND IRON, To clean, from temporary and slight rust.—Cocoanut husks are better than waste and turpentine.

STEEL, CHROME.—This metal is not only one third stronger than any other steel, but can be produced at a small cost, from the fact that when worn out, as in a steel-headed rail, it has a market value, as it can be made over again, which is not the case with Bessemer or any other cast-steel. It will also weld without borax or flux, and when burnt can be redeemed on the next heat.

STEEL, POLISHED, To bronze.—To 1 pint methylated spirits add 4 ozs. gum-shellac and ¼ oz. gum-benzoin ; put the bottle in a warm place, shaking it occasionally. When dissolved and settled, decant the clear liquid and keep it for fine work. Strain the residue through a fine cloth. Take ¼ lb. powdered bronze green, varying to suit the taste with lampblack, red ochre, or yellow ochre. Take as much varnish and bronze-powder as required, and lay it on the article, which must be thoroughly clean and slightly warm. Add another coat if necessary. Touch up with gold-powder according to taste, and varnish over all.

STEEL, Protecting, from rust.—Paraffine is the best material for polished steel or iron.

STEEL RAILS, Cutting.—Remarkable results have been obtained with a disk made from a rail-saw and rotated at 3000 revolutions per minute. As the disk was 9.6 feet in diameter, the velocity of its circumference was in the neighborhood of 86,400 feet per minute. Steel rails were cut with astonishing rapidity, and even melted. Millions of sparks were thrown off, but no heating of the disk could be detected after the cutting.

TIN, Crystallization of.—A platinum capsule is covered with an outer coating of paraffine or wax, leaving the bottom only uncovered. This capsule is set upon a plate of amalgamated zinc in a porcelain capsule. The platinum is then filled completely full of a dilute and not too acid solution of chloride of tin, while the porcelain is filled with water acidulated with $\frac{1}{20}$ of hydrochloric acid, so that its surface comes in contact with the surface of the liquid in the platinum. A feeble electric current is set up, which reduces the salt of tin. The crystals formed after a few days are well developed.. They are washed with water and dried quickly.

TIN, Removing, from copper vessels.—Immerse the articles in a solution of blue vitriol.

TIN, Removing, from plates without acid.—Boil the scrap-tin with soda lye in presence of litharge.

WELDING POWDERS for iron and steel.—(1) Iron filings 1000 parts, borax 500, balsam copaiva, or other resinous oil, 50, sal-ammoniac 75. Mix together, heat, and pulverize. Weld at cherry-red. (2) Borax 15 parts, sal-ammoniac 2, cyanide of potassium 2. These constituents are dissolved in water, and the water itself afterward evaporated at a low temperature.

ZINC, Black color for.—Clean the surface with sand and sulphuric acid, and immerse for an instant in a solution of sulphate of nickel and ammonia 4 parts, in water 40 parts, acidulated with sulphuric acid 1 part. Wash and dry. This takes a bronze color on burnishing.

ZINC LABELS, Ink for writing on.—(1) Verdigris 1 oz., sal-ammoniac 1 oz., lampblack $\frac{1}{2}$ oz., water $\frac{1}{2}$ pint ; mix well in a mortar, and shake before using. Write with a quill. (2) One drachm chloride of platinum dissolved in $\frac{1}{2}$ pint water.

ZINC, Painting.—Use a mordant of chloride of copper 1 part, nitrate of copper 1, sal-ammoniac 1, dissolved in water 64. Add hydrochloric acid (commercial) 1. This brushed over the zinc sheets gives them a deep black color, turning grayish after drying, in from 12 to 24 hours. A coat of oil-color will adhere to this surface and withstand weather excellently.

ZINC-WHITE, To restore.—This may be done by ignition in an earthen crucible.

SIMPLE INSTRUMENTS AND THEIR USES.

BALANCE, Simple spring.—A is a deal stand 12 by 3 inches; B is a hard-wood block firmly attached to A; C is a spring; D is an index-pillar; E is a scale-holder; F is a small bent pin to hold the spring steady while changing the scale-pan. The

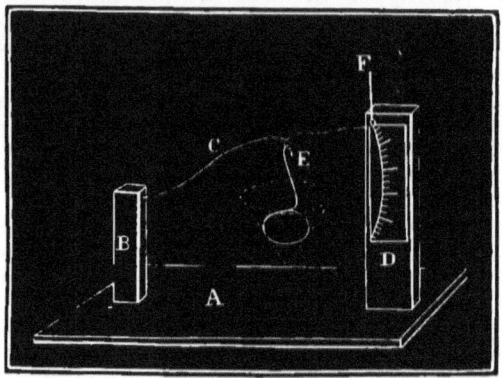

SIMPLE BALANCE.

spring C should be very fine steel wire, bent over so as to form a loop near the index for E to hook into. The index is a slip of card set out with a fine pen. The scale-pan is of thin letter-paper, circular, and folded like a filter-paper, as indicated by the dotted line. With this minute fractions of a grain can be recognized.

BAROMETER, To make a cheap.—Obtain a straight fine glass tube, about 33 inches long, and with as clean an interior as possible, sealed at one end, and having an even uniform bore of about $2\frac{1}{2}$ lines diameter. The mercury to be used should be perfectly pure and free from all air and moisture. This latter requisite may be assured by heating the mercury in a porcelain dish to nearly the boiling-point, previous to using it. The tube is then held securely, with the open end uppermost, and carefully filled with the liquid metal. The open end of the tube is then securely covered with the finger, the tube inverted, and the end covered by the finger plunged below the surface of a little mercury placed in a small vessel to receive it. The finger is then removed, when the mercury in the tube will immediately fall to a level of about 30 inches above the surface of that in the small reservoir below. In order to attach the scale correctly, it will be necessary to compare the indications with those of some good instrument.

BAROSCOPE, To make a.—Take any bottle; pour colored water into it, about $\frac{1}{4}$ of the quantity the bottle will hold; insert in it a glass tube, from 3 to 4 feet long, and passing air-tight through the stopper, which must also be air-tight. Let a paper index, divided according to any scale of division, say into inches and fractions of an inch, be glued to the glass tube. Blow into the glass tube so as to cause the water to ascend the tube a few inches, say 10 inches, and the instrument is constructed. The bottle must be placed in another vessel, and protected by sawdust, or some other material, from the influence of changes in the temperature of the atmosphere. This very sensitive instrument records faithfully any change in the density of the external air, and the approach of a storm will infallibly be indicated by a sudden rise of the water in the glass tube.

CAMERA, WONDER, How to make a.—A wonder camera is a sort of magic lantern, so contrived as to enable one to use opaque objects for projection upon the screen instead of glass transparencies. For example, if a photographer wishes to show his customer how an enlargement from a carte will look, he simply has to put the carte in the wonder camera, and "throw it up." Many enlargement-scales may be made in this way. It consists of a wooden box, with a top made of tin or sheet-iron ; the chimney is made of the same material. The lens is the same as used upon a camera for making photographs. At the back of the box (as will be seen by reference to the elevation and plan, Figs. 2 and 3) are two doors placed upon hinges.

When the box is in use, the door *e* is kept closed. The other door consists of two parts placed at right angles to one another ;

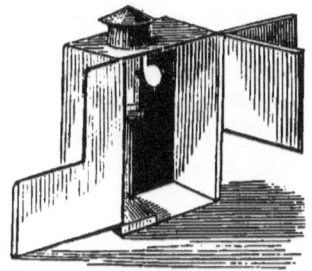

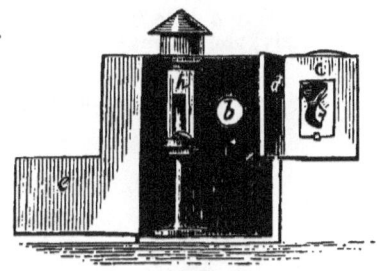

FIG. 1.   FIG. 2.

A WONDER CAMERA.

the object of this is to fill the opening in the door *e* while the pictures are being attached to *c ;* when *c* is swung into position opposite, the lens, placed at *b d*, is carried to one side. If stereoscopic views are to be shown, a slit may be cut at *e*, through which they may be inserted without opening the box. The door *e* should be cut off a little at the bottom, so as to admit air. The light is placed at *h*, as nearly opposite the picture as possible. It should be a strong light ; an argand burner is the best. At the back of the light is a piece of tin, bent into the form of a reflector.

FIG. 3.

The light coming from *h* strikes *c*, and is reflected through the lens upon the screen. The plan of the box is represented with the top removed. No dimensions are given, as they will depend upon the focal distance of the lens and height of

the light. Care must be used to have the distance from the
lens to *c* when closed equal to the focal distance.

ELECTRICAL MACHINE, A simple.—A B, in the annexed en-

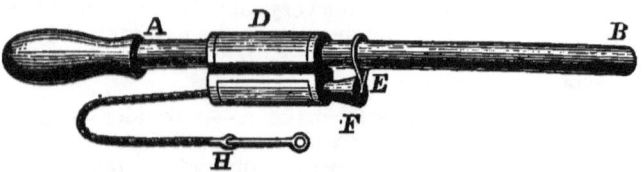

A SIMPLE ELECTRICAL MACHINE.

graving, is a glass tube fixed at one end in a wooden handle.
The rubber, with its flap, D, carries a little Leyden jar, the
end of which is visible at F. This jar is coated inside and
out with a resinous insulating compound, and the metallic lining
of the inside of the jar is in contact with the brass collecting-
ring, E. The handle being held in one hand and the rubber in
the other, when the tube is rubbed the little ring and jar rapidly
collect electricity. A ½ inch spark and smart shock may be readily
obtained from this apparatus, the length of the spark depending
upon the amount of rubbing each time before the jar is discharged. When it is not desired to take the shock through the
human body, the jar may be discharged by means of the metallic
cord, H.

ELECTRICAL ORRERY, to accompany the above machine.—This
is represented below. It is balanced on a pivot at F. The

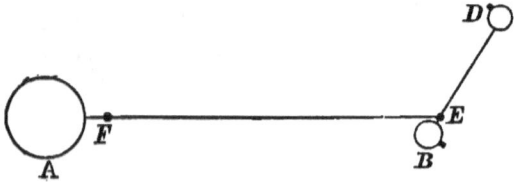

ELECTRICAL ORRERY.

light hollow brass ball, A, represents the sun, and pith balls, B
and D, the earth and moon, rotating about the pivot E. The
metallic points projecting from B and D (in opposite directions,
of course) cause these to rotate round each other ; but the leverage of the point D being, from its position, greater than the
leverage of B, it sets the long arm of the orrery in rotation upon
the pivot F.

GALVANOMETER, To make a simple.—Take an ordinary pocket-
compass and wind 100 feet of No. 18 insulated copper wire
around it.

KALEIDOSCOPE, To make a.—Take two strips of glass, 8 or 10
in. long, 1 to 1½ in. broad at one end and about ½ as broad at the
other. Blacken one side of each with black varnish. Put two
smooth straight edges together, and form a hinge by gluing a
strip of cloth over the two edges. Make the angle between the

strips of glass an aliquot part of 180°, as 20°, 30°, or 45°. Cover the open side of the triangular prism with black velvet. Place in a tin or pasteboard tube so that the angle of the smaller end of prism is nearly in the centre. Cover top of tube with clear glass, and cover this with paper, except a small hole in centre. In bottom of tube, form a cell by placing two pieces of glass ⅛ in. apart (the lower one of ground glass). In this cell place fragments of broken colored glass, beads, etc. They must be capable of free movement in the cell when the tube is turned in the hand.

LEYDEN JAR, To make a cheap.—Line a thin glass candy-jar inside and outside with tin-foil, such as is used to wrap chewing-tobacco in. Stick the foil, on with mucilage, varnish, or flour paste. A still cheaper plan is simply to fill a glass jar nearly full of water, and place it within another vessel of water, so that the water, both outside and inside, shall be on the same level.

MAGIC-LANTERN SLIDES, Painting. *Four methods.*—(1) Use transparent colors, like Prussian blue, gamboge, and carmine. These will give the three primary colors, and by their mixture, the other tints. Apply with a brush, and a transparent drying varnish, like dammar varnish. Allow one coat to dry before applying a second. Considerable aid can be derived from stippling, the color being strengthened, where necessary, by applying it with the point of a fine brush. The colors must not be used too thin. (2) Flow the glass plate with albumen, after the manner of photographers, and paint with aniline colors. This process gives great softness and brilliancy to the pictures, but they are apt to fade. (3) Paint with water-colors, and then flow the entire surface with Canada balsam, covering the painted side with a glass plate. (4) Use water-colors, but mix them with turpentine instead of water, and work rapidly.

MERIDIAN, To find the.—Mr. George W. Blunt says : " Take a piece of board, or any similar material, and describe on it a number of concentric circles. Place this in the sun ; over the centre hang a plummet. Observe the shortest shadow from the plummet ; the sun will then be on the meridian ; draw a line to the centre of the circle, and that will be the true meridian-line. This will do to mark the apparent time, or to correct the compass for variation."

MIRRORS, GLOBE, To make.—Melt together 1 oz. clean lead and 1 oz. of fine tin in a clean iron ladle ; then immediately add 1 oz. bismuth. Skim off the dross, remove the ladle from the fire, and before it sets add 10 ozs. quicksilver ; now stir the whole carefully together, taking care not to breathe over it, as the fumes of mercury are very pernicious. Pour this through an earthen pipe into the glass globe, which turn repeatedly around.

PIPES. Determining proportions of.—The instrument consists simply of a piece of wood shaped like a set-square, as shown in Fig. 1, or a diagram of the same form drawn on paper, and divided out along the two edges, which are at right angles to each other, the divisions being taken to represent inches, feet, or yards, etc., according to the kind of work for which the instrument is used. Suppose that two pipes, A and B, Fig. 2, respectively 5 in.

and 4½ in. in diameter, deliver into a third pipe, D, and it be required to find the proper diameter for the latter pipe. Then from
5 on the scale of one of the divided edges to 4½ on the other,

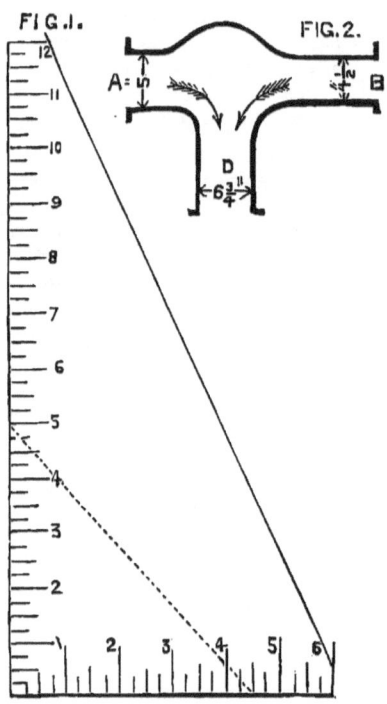

APPARATUS FOR DETERMINING THE DIAMETER OF PIPES.

draw a line, as shown dotted in Fig. 1, and the length of this
line, measured with the same scale as that to which the edges are
divided, will be the diameter of pipe required ; in this case, 6¾
in. On the other hand, if a pipe, D, 6¾ in. in diameter, be delivered into a pipe, A, 5 in. in diameter, and it was required to know
what other size of pipe, B, could also be supplied, all that would be
necessary would be to take the division point 5 on one edge as a
centre, and, with 6¾ in. as a radius, describe an arc cutting the
other divided edge. The point at which the latter edge was cut
by this arc would show the diameter of pipe required.

RIFLE-TELESCOPE, To make a.—Object-glass should be half an
in. in diameter, focus 24 in., or as long as convenient. Eye-piece
may be a single lens of low power with cross spider-lines fixed in
its focus. The target will then appear inverted. The lenses are
inclosed in a brass tube with a hinge or ball-joint at the breech or
eye-piece end, and slides at the muzzle, to depress the object-glass
for increased elevation. The two points of attachment to the barrel are the same as for ordinary fore-and-leaf sights.

TELESCOPE, To make a cheap.—A correspondent says : " I se-

lected a meniscus 1 in. in diameter and of 48-in. focus. This was
for my object-glass. I had already in my possession a two-lensed
double-convex jeweler's eye-glass; one of these lenses was used
for the eye-piece, its focal length being a trifle over 1 in. The
tube was made of pine-wood. A piece of straight, evenly-grained
one-in. pine board, 2 in. wide and 8 feet long, was cut in the middle, and the two pieces, after making a tapering semicircular
groove in each, well glued together. This done, the next thing
was to give it a round, tapering form, 2 in. in diameter at one end,
and a trifle over an inch at the other. This was done with a common carpenter's plane. I now had a tube 4 feet long, with a tapering hole through its length, and 1¼ in. in diameter at its largest end. Two wooden cells for the lenses were then turned in a
lathe, and were made to go on to the tube, as does the cover of a
wooden pill-box. A round hole, the size of the lens, was made
in each, the meniscus being contracted to ¾ in., and the eye-glass
to ¼ in. diameter. The piece carrying the eye-glass was made so
as to slide some distance on the tube, for adjustment to distinct
vision. The tube was painted and varnished, and mounted equa-
torially; and it proved to be a good instrument, showing Jupiter's
moons, their movements and eclipses, handsomely, the ring of
Saturn, the horned appearance of Venus, the mountains and cra-
ters on the moon, the spots on the sun, etc. Several of the nebu-
læ were also visible, especially those in *Andromeda*, *Orion*, *Her-
cules*, and *Sagittarius*. The whole need not cost over two dollars,
beside the time in making, provided one is a mechanic.

"The meniscus (concave on one side and convex on the other)
is the proper form for a single-lens object-glass, and a plano-con-
vex lens makes the best form for the eye-piece. Care must be
taken to so set the lenses in their cells that their foci will meet
centrally. When this is the case, the lenses are said to be well
centred, and in that way we get rid of most of the prismatic
color. Another point that wants attention is the mounting. Absolute steadiness is required for close observation. I used to put
mine upon a post set firmly in the ground. The equatorial arrangement for mounting is described in nearly every work on
telescopes."

THERMOMETERS, Hard-rubber.—This instrument has been made
by riveting the rubber to a thin strip of steel, about a foot in
length and ¼ in. in width. The bottom of this was held fast,
while the top was free to move, and so to indicate the temperature
on a graduated arc. This one, now in use, has a range from zero
to 90° Fahr., and is as sensitive as the common mercurial thermometer. It is well adapted for the ordinary range of the atmosphere, but is not suitable for indicating high degrees of heat,
as the rubber softens at about 200° Fahr. Another thermometer
was made by perforating a thin strip of steel, at intervals of an
inch, and placing upon it a strip of rubber compound when in a
plastic state. This was coiled, with an intermediate strip of metal, which forced the rubber through the holes. It was then vulcanized in the usual manner; and when cold, the intermediate
strip was withdrawn, leaving an open space between the coils.
This saved the trouble of riveting, and gave to the rubber an un-
broken and smooth surface. The coil is held fast at the centre,

and the outer end is left free to move. Another thermometer was made of glass and hard rubber, the latter in the form of an arc, being riveted at both of its ends to a glass plate, which formed the chord.

THERMOMETERS.—To change Fahrenheit degrees into Centigrade: $C.=\dfrac{5\,(F.-32)}{9}$. Centigrade into Fahrenheit: $F.=\dfrac{9\,C.}{5}+32$.

Réaumur into Fahrenheit: $F.=\dfrac{9\,R.}{4}+32$. Fahrenheit into Réaumur: $R.=\dfrac{4\,(F.-32)}{9}$. Réaumur into Centigrade: $C.=\dfrac{5\,R.}{4}$.

Centigrade into Réaumur: $R.=\dfrac{4\,C.}{5}$

RECIPES FOR THE PREPARATION OF WOOD.

DYEING WOODS.—All light woods may be died by immersion. A fine crimson is made as follows : Take 1 lb. ground Brazil, boil in 3 quarts of water, add $\frac{1}{2}$ oz. cochineal, and boil another half hour ; may be improved by washing the wood previously with $\frac{1}{2}$ oz. saffron to 1 quart water. The wood should be pear wood or sycamore. Purple satin : 1 lb. logwood chips, soak in 3 quarts of water, boil well an hour ; add 4 ozs. pearlash, 2 ozs. powdered indigo. Black may be produced by copperas and nutgalls, or by japanning with two coats of black japan, after which varnish or polish, or use size and lampblack previous to laying on japan. A blue stain : 1 lb. oil of vitriol put in a glass bottle with 4 ozs. indigo ; lay on the same as black. A fine green : 3 pints of the strongest vinegar, 4 ozs. best powdered verdigris (poison), $\frac{1}{2}$ oz. sap-green, $\frac{1}{2}$ oz. indigo. A bright yellow may be stained with aloe ; the whole may be varnished or polished.

EBONY, Artificial.—Treat sea-weed for 2 hours in dilute sulphuric acid. Of the charcoal thus obtained take 16 parts ; dry, and grind it. Add liquid glue 10 parts, gutta-percha 5, india-rubber $2\frac{1}{2}$, the last two dissolved in naphtha. Then add coal-tar 10, pulverized sulphur 5, pulverized alum 2, powdered resin 5, and heat the mixture to 300° Fahr. This when hard will take a polish equal to ebony, and is the same in color and hardness.

OAK, To color orange-yellow.—Rub the wood with a mixture of tallow 3 ozs., wax $\frac{3}{4}$ oz., and turpentine 1 pint, mixed by heating together and stirring. Apply in a warm room until a dull polish is acquired. Then coat, after an hour, with thin polish, and repeat until the desired depth and brilliancy of tone is obtained.

SCREWS, WOODEN, To season.—Bore a hole longitudinally through the centre of the screw ; it will not be apt to crack so badly in seasoning, because then the air can get to the centre of the wood, the sap escapes therefrom, the centre of the wood contracts, and the strain on the outside is lessened. Of course, the larger the hole, the better for the seasoning process ; but it should not, and need not, be large enough to materially weaken the screw. If, in addition, you can boil the screw in water, the job will be bettered ; if boiled in oil, it will be complete.

VENEERS, ARTIFICIAL, To make.—Soak the wood for 24 hours, and boil for $\frac{1}{2}$ hour in a 10 per cent solution of caustic soda. Then wash out the alkali, when the wood will be elastic, leather-like, and ready to absorb the desired color. After immersion in the color-bath, dry between sheets of paper under sufficient pressure to preserve the shape.

VENEERS, Steaming.—Blocks of wood intended for veneers may be steamed in a solution of borax and ammonia. They will then become soft and easy to cut, and, beside, will retain their flexibility for a long time.

WOOD, A liquid.—Sawdust can be converted into a liquid wood, and afterward into a solid, flexible, and almost indestructible mass, which, when incorporated with animal matter, rolled, and dried, can be used for the most delicate impressions, as well as for the formation of solid and durable articles, in the following manner : Immerse the dust of any kind of wood in diluted sulphuric acid, sufficiently strong to affect the fibres, for some days ; the finer parts are then passed through a sieve, well stirred, and allowed to settle. Drain the liquid from the sediment, and mix the latter with a proportionate quantity of animal offal, similar to that used for glue. Roll the mass, pack it in moulds, and allow it to dry.

WOODEN LABELS, Preservation of.—Thoroughly soak the pieces of wood in a strong solution of sulphate of iron ; then lay them, after they are dry, in lime-water. This causes the formation of sulphate of lime, a very insoluble salt, in the wood. The rapid destruction of the labels by the weather is thus prevented. Bast, mats, twine, and other substances used in tying or covering up trees and plants, when treated in the same manner are similarly preserved. Wooden labels, thus treated, have been constantly exposed to the weather during two years without being affected thereby.

WOODEN TAPS, Preserving, for Casks.—The articles should be plunged in paraffine, heated to about 248° Fahr. until no air-bubbles rise to the surface of the melted material. They are then allowed to cool, and the paraffine is removed from the surface, when nearly congealed, by thorough rubbing. Taps thus treated will never split or become impregnated with the liquid, and may be used in casks containing alcoholic liquors.

WOOD, Brown stain for.—Paint the wood over with a solution made by boiling 1 pint catechu (cutch or gambier) with 30 pints water and a little soda. Dry, and then paint over with a solution of bichromate of potash 1 pint, water 30 pints. By a little difference in the mode of treatment, and by varying the strength of the solutions, various shades of color may be given to these materials. The colors will be permanent, and will tend to preserve the wood.

Wood, Preserving.—This process is valuable for railway-sleepers. Steam the timber, and inject a solution of silicate of soda for 8 hours. Then soak the wood for the same period in lime-water. (Dr. Feuchtwanger's process.)

Wood, Preservative preparation for.—Mix 40 parts chalk, 50 resin, 4 linseed-oil, melting them together in an iron pot ; then add 1 part of native oxide of copper, and afterward 1 part of sulphuric acid. Apply with a brush. When dry, this varnish is as hard as stone.

Wood, To ebonize.—Collect lampblack from a lamp or candle on a piece of slate. Scrape off the deposit, mix with French polish, and apply to the object in the ordinary way.

Wood, To fire-proof.—Paint twice over with a hot saturated solution of 1 part green vitriol and 3 parts alum. After drying, paint again with a weak solution of green vitriol in which pipe-clay has been mixed to the consistence of paint.

THE PREPARATION AND PRESERVATION OF NATURAL-HISTORY SPECIMENS.

Anatomical Specimens, Preserving.—Glycerine will preserve the natural colors of marine animals kept immersed in it.

Birds, Stuffing.—The following tools are required (see Fig. 1). —First, there is the scalpel. This can be purchased for a small sum from any maker of surgeon's instruments. The blade is short and very sharp, while the handle (not jointed) is long enough to allow of a firm grasp. From the same maker, a couple of pairs of surgeon's scissors should also be obtained, one quite small and sharp-pointed, the other of medium size ; also two or three spring forceps of various dimensions. A small pair of pliers for clipping wire is required, some spools of cotton (Nos. 10, 30, and 100), a quantity of excelsior and tow, some cotton batting, a little prepared glue, a number of pieces of wire about fifteen inches long, and straight (size No. 20 or thereabouts), a box of dry oatmeal, and some arsenical soap. This last can generally be obtained of druggists, or, if not, can be made of carbonate of potash 3 ounces, white arsenic, white soap, and air-slaked lime, 1 ounce each, and powdered camphor, $\frac{3}{16}$ of an ounce. This is combined into a thick paste with water, and applied as below described, with a small paint-brush. It should be marked as POISON, and kept scrupulously out of the reach of children or pet animals.

If the bird has been shot, immediately afterward all the holes made in its body, as well as the mouth, should be plugged with cotton, in order to prevent the escape of blood or liquids. Operations should not be begun for twenty-four hours, so that the body may have ample time to stiffen and the blood to coagulate. It is well during this period to inclose the bird, head downward, in a cone of paper, so that the feathers will be held smooth.

The first process is skinning. In commencing, the left hand is used to part the feathers, exposing the skin from the apex of the breast-bone to the tail. With the scalpel held like a pen, a free incision is made between these points, care being taken to divide the skin only, without cutting into the flesh. The skin is then pressed apart, and oatmeal dusted into the cut, in order to absorb any fluids which may escape. Careful lifting of the skin clear of the flesh follows, until the leg is reached, when the scalpel is again used to disarticulate the thigh-joints. The bone of each thigh is then exposed for its whole length, by pushing back the skin, and the meat removed, when the bone is replaced, and the other thigh treated in similar manner.

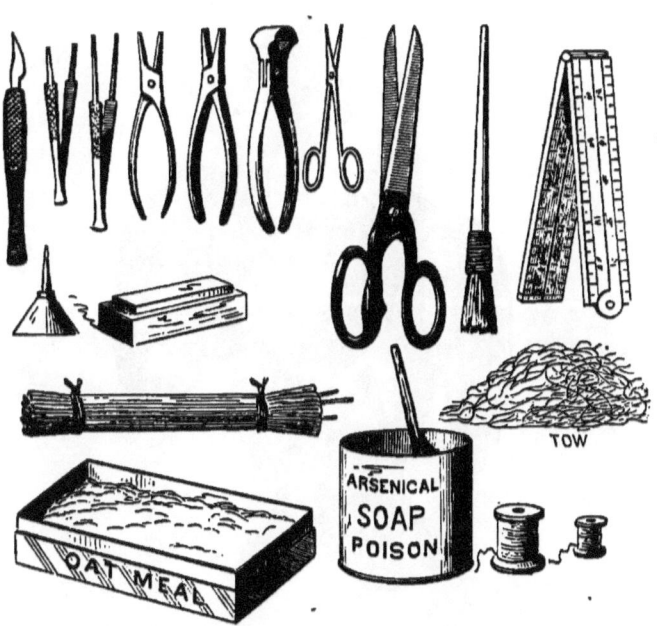

FIG. 1.—TAXIDERMICAL IMPLEMENTS.

The skin is next detached, to the wings, which are cut from the body at the joint next the same, and the bones scraped clear of meat. Then the neck is divided, so that the skin, with the head attached, can be peeled from the entire body clear to the root of the tail. The last is bent toward the back with the left hand, the finger and thumb keeping down the detached parts of the skin on each side of the vent. A deep cut is then made across the latter until the back-bone, near the oil-gland at the root of the tail, is exposed. Sever the back-bone at the joint. This detaches the body, which may be removed and thrown aside, while the root of the tail, with the oil-gland, is left. Great care is needed in this operation, as, if not enough bone be left at its root, the tail will come out ; but all fleshy matter should be neatly dissected away.

The neck need not be split or in any wise cut. The skin is merely pulled over the flesh, as a glove is removed from the finger, until the skull is exposed and appears as in the sketch, Fig. 2. With the point of the knife, remove the ears ; and on reaching the eyes, carefully separate the lids from the eyeballs, cutting neither. It requires very delicate and slow work at this point, so as not to injure the eyelids. Then scrape out the eye cavities, and cut away the flesh of the neck, removing at the same time a small portion of the base of the skull. Through the cavity thus made extract the tongue and brains, and after cleaning away all fleshy matter, paint the eye orbits with arseni-

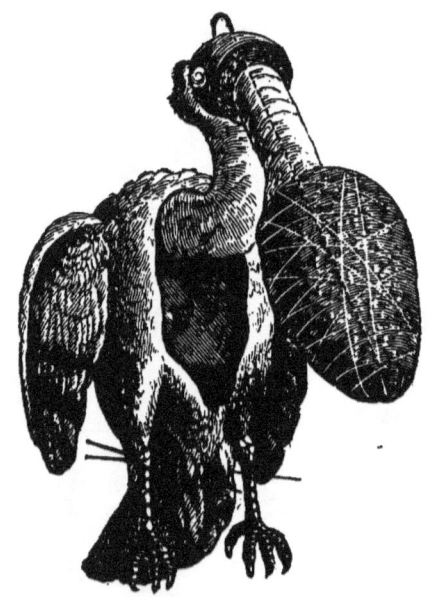

FIG. 2.—MODE OF ATTACHING THE FALSE BODY.

cal soap, and stuff them tightly with cotton. Care should be taken not to detach the skin from the bill, as it is necessary to leave the skull in place. Finally, fill the interior of the skull with tow (never with cotton), after coating internally with the prepared soap.

The skinning operation being now completed, the stuffing is next proceeded with. To prepare for this, the bird, before being skinned, should have been measured, first as to its girth about the body, and second as to its length from root of tail to top of skull, following the shape of the form. From these data an artificial body of the right dimensions is constructed and inserted as follows : On a piece of straight wire, equal in length to the last measurement above mentioned, a bunch of excelsior is secured by repeated winding with stout thread. This bundle, which is represented in our Fig. 2, is moulded to a shape resembling that of the bird's body, and its girth is regulated by the similar measurement already obtained from the bird itself. As will be seen, it is attached at the end of the wire, the long protruding portion of which serves as a foundation for the neck. The extremity of the wire is clipped by the pliers to a sharp point, and then forced diagonally upward through the skull, on top of which it is clinched flat. Cotton batting is then wound about the wire between skull and body, until sufficient thickness is obtained to fill the skin of the neck. The position of the various parts at this point is represented in Fig. 2. Painting the inside of the skin with arsenical soap follows, and then the skin is drawn back so as to envelop the false body, and a needle and thread are thrust through the nostrils to make a loop for convenience in handling.

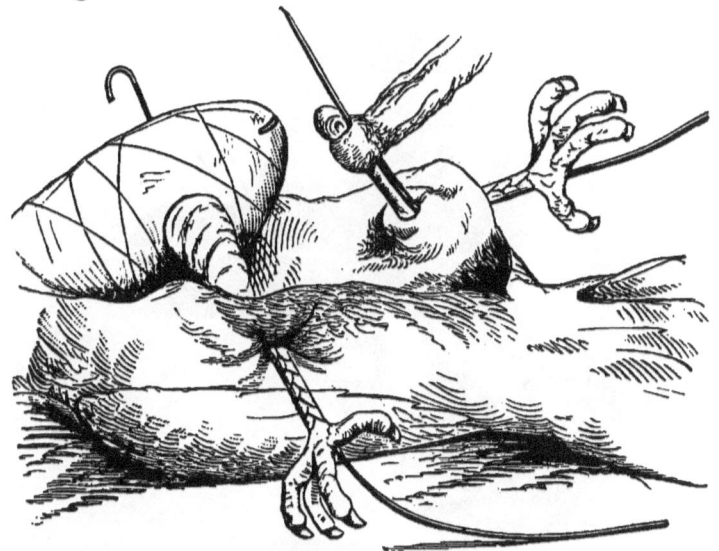

FIG. 3.—STUFFING THE LEGS.

The finest pair of forceps is employed to pull the eyelid skin into place, to arrange the feathers, and to pull up the cotton in the orbits so as to stuff the cavities out plumply. More cotton is next pushed down the throat until the same is entirely filled. Two pieces of wire—quite stout for large bird—are then sharpened at one extremity. Taking the wire in one hand and guiding it with the other, the operator shoves it into the leg, from the ball of the foot up alongside the thigh bone, the skin being turned back for the purpose. Cotton is then wound about both wire and bone, in order to fill the thigh out naturally, and the same process is repeated for the other side. The ends of the wire below are left protruding in order to support the bird on a perch, if such be desired. The upper ends are pushed clean through the artificial body, from below up, and clinched on the upper side. This secures the legs, which are afterward bent in natural position (Fig. 3). The bird can now be set up—that is, the wires stretching out below the claws can be wound about a perch or pushed through holes in a board and clinched on the under side. In the latter case, it will be necessary to spread the claws and fasten them with pins. For small birds, the cut in the breast need not be sewn up ; a chicken or larger fowl will require a few stitches to hold the edges together. If the tail feathers are to be spread, a wire is thrust across the body and through each feather, holding all in the proper position. The wings are then gathered closely into the body, and two wires,

FIG. 4.—THE BIRD PREPARED FOR DRYING.

one from each side, are pushed in diagonally from up, down, and through the skin of the second joint (Fig. 4). The wings are thus held, and the wires, as well as that through the tail, are left protruding for an inch or more. A touch of glue within the eyelids prepares the latter for the eyes. These must be purchased from taxidermists, but for small birds common black beads will answer. If plain glass beads can be obtained, by the aid of a little paint the student can easily imitate the eye of a chicken. After the eyes are inserted, a sharp needle is used to pull the lids around them and into place.

The operator must now, with a fine pair of forceps, carefully adjust the feathers, smoothing them down with a large camel's-hair brush. This done, thread must be wound over the body very loosely, beginning at the head, and continuing until all the feathers are securely bound. The bird is then left to dry for a day or two, when the thread is removed, the ends of wire cut off close to the body, and the work is complete.

ENTOMOLOGICAL SPECIMENS, To preserve, from insect ravages. —Place crystals of carbolic acid throughout the cabinets, and the evaporation of the crystals will keep them thoroughly saturated with carbolic acid vapors, which will kill all living insects therein.

FISH, To mount and preserve.—It is impossible to preserve the iridescent tints of the living specimens; but before proceeding to the operation of skinning, it may be stated that the scales, as well as their color, may be preserved to a certain degree by applying tissue-paper to them, which, from the natural glutinous matter which covers the scales, will adhere firmly; this being allowed to remain until the skin has dried, may easily be removed by moistening with a damp cloth. All small fish should be mounted in section, while the larger varieties may be preserved entire. Supposing the fish to be of such a size as to be mounted in section, first it is necessary that it be as fresh as possible, as the scales will become detached if decay set in. Place the fish on one side, and cover the side uppermost with tissue-paper, as above stated; also extend the fins by means of the same, and allow them to remain a few moments until fixed and dry. Having provided yourself with a damp cloth, spread it smoothly upon the table, and place the fish upon it, papered side down. With the dissecting scissors, cut the skin along an oval line, following the contour of the body, but a little below the extreme dorsal edge and a little above the ventral one, and remove the skin included within this line. The remaining skin must now be detached from the flesh, beginning at the head and separating it downward toward the tail. The spine must be severed close to the head, and also at the tail, and the entire body removed. All the flesh having been taken from the skin, and the eyes removed, the inside must be wiped out and the preservative (arsenical soap) applied. The skin should now be filled with tow, very evenly placed. When filled, it should be laid, with the open side down, upon a board of proper dimensions, and fastened to it by small tacks, beginning at the head and fastening the edges downward toward the tail. It should then be set aside to dry. The paper is, after drying, removed, and eyes of wood (painted to the proper colors, and not varnished) are inserted with a little putty. Finally the skin should receive a coat of colorless varnish, when the specimen is ready for the cabinet.

SEA-WEED, Preserving specimens of.—The best time to collect is when the tide has just commenced to flow, after the lowest ebb, as the sea-weeds are then floated in, in good condition. All specimens should be either red, green, purple, black, or olive; no others are worth preservation.

Mounting is done by immersing a piece of paper just below the surface of the water, and supporting it by the left hand; the weed is then placed on the paper and kept in its place by the left thumb, while the right hand is employed in spreading out the branches with a bone knitting-needle or a camel's-hair pencil. If the branches are too numerous, which will be readily ascertained by lifting the specimen out of the water for a moment, pruning should be freely resorted to, by cutting off erect and alternate branches, by means of a sharp-pointed pair of scissors, close to their junction with the main stem. When the specimen is laid out, the paper should be raised gradually in a slightly sloping direction, care being taken to prevent the branches from running together. The delicate species are much improved in appearance by reimmersing their extremities before entirely withdrawing them from the water. The papers should then be laid flat upon coarse bibulous paper, only long enough to absorb superfluous moisture. If placed in an oblique direction, the branches are liable to run together. They should be then removed and placed upon a sheet of thick white blotting-paper, and a piece of washed and pressed calico placed over each specimen, and then another layer of thin blotting-paper above the calico. Several of these layers are pressed in the ordinary way, light pressure only being used at first. The papers, but not the calico, may be removed in six hours, and afterward changed every 24 hours until dry. If the calico be not washed, it frequently adheres to the algæ, and if the calico be wrinkled it produces corresponding marks on the paper. The most convenient sizes of paper to use are those made by cutting a sheet of paper, of demy size, into 16, 12, or 4 equal pieces. Ordinary drawing-paper answers the purpose very well. For the herbarium, each species should be mounted on a separate sheet of demy or cartridge size. Toned paper shows off the specimens well, a neutral tint answering best for the olive, pink for the red, and green for the green series.

SKINS OF SMALL ANIMALS, Dyeing.—The green hull of the European walnut is turned to account in Europe for dyeing furs black, and the hull of our black walnut could probably be similarly employed. The walnut hull is crushed and the juice squeezed out from the pulp, with the addition of a little water. A small quantity of lime is added, and the dye is ready for use. The color is extremely difficult of extraction, and attaches itself very readily to any kind of hair, and it is used extensively as a hair-dye.

STUFFING SMALL QUADRUPEDS.—Begin by making a longitudinal incision between the hind legs, extending quite back to the vent, the hair having been carefully parted so that it may not be cut. Do not cut into the abdominal cavity. The skin can now be separated from the flesh and turned back as far as the thigh, which is severed at the joint. When this is done on both sides, the gut should be drawn out and severed at a short distance from the vent. The tail should also be disjointed at the root. This being done, the skin can be loosened around the body until the fore-legs are reached, when they also should be dissevered. The skinning now proceeds along the neck until the skull is reached. Here considerable care is necessary to remove the skin without damage to the ears, eyelids, and lips.

The skin is left attached to the skull; when the operation has proceeded far enough to expose the muscles of the jaws, the skin must be separated from the body at the first joint of the neck. The tongue, eyes, and muscles, remaining attached to the head, are now to be carefully removed, and the brain taken out from an opening in the back of the skull cut through for that purpose. To make this opening, amateurs can use a small gimlet or bit with very small animals, and a large one as circumstances may demand. The legs are now to be skinned out quite down to the claws, which completes the operation of skinning. During the entire process, all fluids escaping must be immediately soaked up with cotton. As soon as the skin is removed, it should be thoroughly rubbed with arsenical soap, not omitting the inside of the skull and mouth cavities.

The following explanation of stuffing relates to a small animal such as the squirrel. Provide yourself with cotton, thread, and twine, also stuffing-forceps, a pair of pincers, a file, and wire-cutters. With the aid of the forceps (a pair of slender-jawed pliers), supply the various muscles of the face and head by inserting cotton both through the mouth and eyelids. Take annealed iron wire and cut off 6 pieces: No. 1, two or three inches longer than the total length of the body; Nos. 2 and 3 for the fore-legs; Nos. 4 and 5 for the hind-legs; each of these should be three inches longer than the limbs they are to support; No. 6, for a support for the tail, of the same proportionate length as the others. With a large pair of scissors, cut fine a quantity of tow, and with this, by the use of the long forceps, stuff the neck to its natural dimensions. Taking wire No. 1, bend it in four small rings, the distance between the two outer representing the length of the body taken from the skin, *a*, leaving one long end for a support to the head and neck, *b* (see

STUFFING ANIMA

figure). Mould tow about that part containing the rings, and, by winding it down with thread, form an artificial body. Sharpen the projecting end of the wire to a fine point with the file, and insert it up through the tow in the neck, and thence through the skull; the skin should then be pulled over the body. Wires No. 2 and 3 are placed next in position by inserting them through the soles of the feet, up within the skin of the leg, and through the body of tow, until they appear upon the opposite side. With the pincers, bend over the end of each, forming a hook; the wires must then be pulled backward, thus fastening the hooks firmly into the body. The loose skin of the limbs should then be stuffed with cut tow, taking care to imitate the muscles of the living subject. Nos. 4 and 5 can be fixed in position after the same manner, unless the animal is to rest entirely upon its rear (as in the case with the squirrel feeding) ; then the wire must be inserted at the tarsal joint instead of at the sole of the foot. If any depressions appear in the skin, they must be stuffed out with cut tow. Wire No. 6 should now be inserted at the tip of the tail, and forced down within the skin, hooking it into the body in the same manner as the leg-wires. Stuff the tail to its proper dimensions with cut tow, and carefully sew up the incision along the abdomen. Having prepared a board about ¼ inch thick, pierce in it two holes at the proper distance apart for the reception of the wires (four holes will be needed if the animal is to stand on all extremities) ; these must be drawn through upon the under side until the feet rest close upon the upper surface, when they should be clinched. The different joints of the limbs can now be imitated by bending the wire at the proper points. The eyes should next be placed in position, and cemented in the orbits by a little putty. Care should be taken in arranging the eyelid, for the expression depends altogether upon this point. Clip off any superfluous wire which may extend above the head with the wire-cutters. The specimen should be placed in some locality free from moisture, and allowed to dry thoroughly, when it is complete for the cabinet.

PAINTING, GILDING, AND VARNISHING RECIPES.

BALLOON VARNISHES.—Mr. John Wise, the well-known aeronaut, says : " There are two ways of preparing linseed-oil for balloon varnish : the quick and the slow process. The first is by heating the oil up to a temperature at which it will ignite spontaneously. In order to secure it from burning up, it must be heated in an iron or copper vessel, with a lid that can be closed when it begins to emit dense white vapor. If it is desired to have it fast drying, from 4 to 6 ozs. litharge per gallon should be boiled in it. This process takes about one hour, and renders the oil thick and tough, giving a good body and glossy surface to the cloth. The slow process is to boil the oil from 12 to 20 hours, keeping it at a temperature of about 200° Fahr., incorporating with it, while boiling, ½ oz. sulphate of manganese to each gallon of oil. These varnishes should be applied to the cloth tolerably hot. There are other formulas, such as the incorporation with the oil of some birdlime. Gum-elastic is also used to give the oil body and elasticity. When I desire to make a balloon extraordinarily close, I give it a first coating of compound varnish made up of equal parts white glue and glycerine."

BRONZE, GOLD, for furniture.—Gold bronze for furniture is a mixture of copal varnish mixed with gold-colored bronze-powder. The last is bisulphate of tin.

BRUSHES, Care of varnishing.—A good way to keep brushes

is to suspend them by the handles in a covered can, keeping the points at least half an inch from the bottom, and apart from each other. The can should be filled with slowly-drying varnish up to a line about $\frac{1}{16}$ inch above the bristles or hair. The can should then be kept in a close cupboard, or in a box fitted for the purpose. As wiping a brush on a sharp edge will gradually split the bristles and cause them to curl backward, and eventually ruin the brush, the top of the can should have a wire soldered along the edge of the tin, turned over, in order to prevent injury. Finishing brushes should not be cleansed in turpentine, except in extreme cases. When taken from the can, prepare them for use by working them out in varnish; and before replacing them, cleanse the handles and binding in turpentine.

COLORS, Naturally transparent.—These are terra de sienna, asphaltum, dragon's blood, carmine, rose-pink, chemical brown, all the lakes, gamboge, and all the gums. Semi-transparent: umber, Vandyke brown, chrome red, emerald green, Brunswick green, ultramarine, indigo, and verdigris. Transparent colors are purer if ground in water; allow them to settle, pour off the top part of the settlings; mix that with more water; let it settle, and take the top half of that, which will be free from all sand and grit. Turpentine makes transparent colors work crumbly. Bleached boiled oil or white varnish is the best vehicle for flowing evenly.

GILDING WITHOUT A BATTERY.—Dissolve 20 grains chloride of gold in a solution of cyanide of potassium, 1 oz. to 1 pint pure water. Put the solution of cyanide of gold in a glass or porcelain jar; place in it the articles to be gilded in contact with a piece of bright zinc, in the solution near them; the process will be hastened by a gentle warmth. If the gold is deposited on the zinc, rub a little shellac-varnish on it. The chloride of gold may be prepared by dissolving gold in *aqua regia* in the proportions of 16 grains gold to 1 oz. acid, and evaporating to dryness.

GILDING ON GLASS.—Mix powdered gold with thick gum-arabic and powdered borax; with this trace the design on the glass, and then bake it in a hot oven. The gum is thus burned and the borax vitrified, and at the same time the gold is fixed on the glass. To make powdered gold, rub down gold-leaf with pure honey on a marble slab. Wash the mixture, and the precipitate is the gold used.

JAPAN, Black and flexible.—Take burnt umber 4 ozs., asphaltum 2 ozs., boiled oil 2 qts.; dissolve the asphaltum first in a little oil, using moderate heat; then add the umber (ground in oil), and lastly the rest of the oil, and incorporate thoroughly. Thin with turpentine.

LOOM-HARNESS, Varnish for.—Mix linseed-oil 2 gals., gum-shellac 2½ lbs., litharge 2 lbs., red-lead 1 lb., umber 1½ lbs., sugar of lead 1½ lbs.

MACHINERY, Painting.—The following colors contrast handsomely: 1. Black and warm brown. 2. Violet and pale green. 3. Violet and light rose-color. 4. Deep blue and golden brown. 5. Chocolate and bright blue. 6. Deep red and gray. 7. Maroon and warm green. 8. Deep blue and pink. 9. Chocolate and pea-green. 10. Maroon and deep blue. 11. Claret and buff. 12. Black and warm green.

MARBLE, To stain.—Blue, solution of litmus ; green, wax colored with verdigris ; yellow, tincture of gamboge or turmeric ; red, tincture of alkanet or dragon's blood ; crimson, alkanet in turpentine ; flesh, wax tinged with turpentine ; brown, tincture of logwood ; gold, equal parts of verdigris, sal-ammoniac, and sulphate of zinc in fine powder.

PAINT without oil.—Break an egg into a dish and beat slightly. Use the white only, if for white paint ; then stir in coloring matter to suit. Red-lead makes a good red paint. To thin it, use a little skimmed milk. Eggs that are a little too old to eat will do for this very well.

PAINT, Reddish-brown, for wood.—The wood is first washed with a solution of 1 lb. cupric sulphate in 1 gallon water, and then with ¼ lb. potassium ferrocyanide dissolved in 1 gallon water. The resulting brown cupric ferrocyanide withstands the weather, and is not attacked by insects. It may be covered, if desired, with a coat of linseed-oil varnish.

PAINT to stand the action of hot water.—Clean the metal with turpentine or benzine. Then mix white-lead, carriage-varnish, and spirits of turpentine, and give the metal two thin coats, and then a thick coat of white-lead and carriage-varnish, applied as quickly as possible.

PUTTY, Indestructible.—Boil 4 lbs. brown umber in 7 lbs. linseed-oil for 2 hours ; stir in 2 ozs. wax ; take from the fire, and mix in 5½ lbs. chalk and 11 lbs. white-lead, and incorporate thoroughly.

IRON SURFACES, Painting.—In mixing paints for iron surfaces, it is of the first importance that the best materials only be used. Linseed-oil is the best medium, when free from admixture with turpentine. A volatile oil, like turpentine, can not be used with advantage on a non-absorbent surface like that of iron, for the reason that it leaves the paint a dry scale on the outside, which, having no cohesion, can be readily crumbled or washed away. Linseed-oil, on the other hand, is peculiarly well adapted for this purpose. It does not evaporate in any perceptible degree, but the large percentage of linolein which it contains combines with the oxygen of the air, and forms a solid, translucent substance, of resinous appearance, which possesses much toughness and elasticity, and will not crack or blister by reason of the expansion and contraction of the iron with variations of temperature. It is, moreover, remarkably adhesive, is impervious to water, and is very difficult of solution in essential oils, spirits, or naphtha, and even in bisulphide of carbon. Another important advantage of linolein is that it expands in drying, which peculiarly adapts it to iron surfaces ; since cracks, however minute, resulting from shrinkage, expose enough of the metal to afford a chance for corrosion, which will spread in all directions, undermining the paint and causing it to scale off, beside discoloring it. In selecting a paint for iron, mechanical adhesion is a consideration of the first importance. Pitchy or bituminous films are especially effective as regards their adhesion to iron ; for example, solutions of as-phalt or pitch in petroleum or turpentine. These are also very effective as regards continuity, owing to the fact that, in drying, they form plastic films, which yield with the expansion and contraction of the iron, and manifest no tendency to crack. If the surface is rusty, they penetrate the oxide scale, and envelop the particles very effectually, making them a portion of the paint. The solubility of such a film in water may be counteracted by mixing it with linseed-oil. The experiment may easily be tried by mixing about 2 parts of Brunswick black with one of white, red, or stone colored paint, the body of which is composed of red or white lead or litharge. Red-lead is the best, for many reasons, if finely ground and thoroughly mixed with linseed-oil. Any one of several kinds of bitumen may be used, either natural mineral asphalt, pine pitch, or artificial asphalt, such as gas-tar or the residuum of petroleum distillation, in cases where the crude oil has been distilled before being treated with acid. This gives a very hard, bright pitch, which is soluble in "once run" paraffine spirit, and which makes the base of an excellent, cheap, and durable paint for iron-work in exposed positions. Paraffine can be recommended for all classes of iron-work which can be treated hot. The most effective method of applying it is to heat the iron in vacuo, in order to expand it and open its pores, when paraffine, raised to the proper temperature, is run upon it. By this means the iron is penetrated to a sufficient depth to afford a very effectual protection against oxidation, especially when a suitable paint is subsequently applied.

LACQUER, Deep golden.—Seed-lac 3 oz., turmeric 1 oz., dragon's blood ¼ oz., alcohol 1 pt. Digest for a week, frequently shaking. Decant and filter. *Golden :* Ground turmeric 1 lb., gamboge 1¼ ozs., gum-sandarac 3½ lbs., shellac ¾ lb. (all in powder), rectified spirits of wine, 2 gals. Dissolve, strain, and add 1 pt. of turpentine varnish. *Red :* Spanish anatto 3 lbs., dragon's blood 1 lb., gum-sandarac 3½ lbs., rectified spirits 2 gals., turpentine varnish 1 qt. Dissolve, strain, and mix, as last. *Pale brass :* Gamboge, cut small, 1 oz., Cape aloes, ditto, 3 ozs., pale shellac 1 lb., rectified spirits 2 gals. Dissolve and mix as with the golden. *Another golden :* Ground turmeric 1 lb., gamboge, powdered, 1½ ozs., gum-sandarac, powdered, 3½ ozs., shellac ¾ lb., spirits of wine 2 gals. After being agitated, dissolved, and strained, add 1 pt. of turpentine varnish, well mixed. *Another red :* Spirits of wine 2 gals., dragon's blood 1 lb., Spanish anatto 3 lbs., gum-sandarac 4½ lbs., turpentine 2 pts. Make in same manner. *Another pale brass :* Spirits of wine 2 gals., Cape aloes, cut small, 3 ozs., pale shellac 1 lb., gamboge, cut small, 1 oz. ; no turpentine varnish. Make in same manner. *Another deep golden :* Strongest alcohol 4 ozs., Spanish anatto 8 grains., powdered turmeric 2 drachms, red-saunders 12 grains. Infuse this mixture in the cold for forty-eight hours, pour off the clear, and strain the rest ; then add powdered shellac ½ oz., sandarac 1 drachm, mastic 1 drachm, Canada balsam 1 drachm. Dissolve in the cold by frequent agitation, laying the bottle on its side to present a greater surface to the alcohol. When dissolved, add 30 drops of spirits of turpentine. *Pale tin :* Strongest alcohol 4 ozs., powdered turmeric 2 drachms, hay saffron 1 scruple, dragon's blood, in powder, 2 scruples, red saunders ¼ scruple. Infuse, and add shellac, etc., as to the last-described deep golden. When dissolved, add 40 drops of spirits of turpentine. Lacquer should always stand till it is quite fine before it is used.

LETTERING, Sign-painter's scale for.—The following is a convenient table for sign-painters or others who have occasion to make lettering. Supposing the height of the capital letters to be ten, the widths are as follows : B, F, P, ten ; A, C, D, E, G, H, K, N, O, Q, R, T, V, X, and Y, eleven ; I, five ; J, eight ; S and L, nine ; M and W, seventeen ; Z and &, twelve. Numerals : 1 equals five ; 2, 3, 5, 7, 8, nine ; 4, eleven ; 6, 9, 0, ten. Lowercase letters (height six and a half): Width : a, b, d, k, p, q, x and z, seven and a half ; c, e, o, s, seven ; f, i, j, l, t, three ; g, h, n, u, eight ; m, thirteen ; r, v, y, six ; w, ten.

PUTTY, Old, in sashes, To soften.—Run a red-hot iron over it : it will peel off easily.

VARNISH, Black.—Alcohol 1 qt., aniline blue 184.8 grs., fuchsin 46.2 grs., naphthaline yellow 123.2. Dissolve by agitation in less than 12 hours. One application is sufficient. The mixture should be filtered when it will not deposit.

VARNISH, Cheap gold.—The following is a cheap substitute for the expensive gold varnish used on ornamental tin-ware. Turpentine ½ gallon, asphaltum ½ gill, yellow aniline 2 ozs., umber 4 ozs., turpentine varnish 1 gal., and gamboge ¼ lb. Mix and boil for ten hours.

VARNISH, COPAL, To make.—Dissolve 1 pt. camphor, by weight, in 12 pts. ether, then add best copal resin (pulverized) 4 pts., and place in a well-stoppered bottle. When the copal has partly dissolved and has become swollen, add strong alcohol 4 pts., oil of turpentine ½ pt. Shake, and allow to stand for a few hours. This makes an excellent varnish.

VARNISH FOR MAPS.—Take equal parts genuine Canada balsam and oil of turpentine ; mix. Set the bottle in warm water, and agitate until the solution is perfect ; then set in a warm place a week to settle, when pour off the clear varnish for use. Before using, cover the map with a thin solution of pure glue.

VARNISH, Parisian.—Dissolve 1 part of shellac in 3 to 4 parts of alcohol of 92 per cent in a water-bath, and add cautiously-distilled water, until a curdy mass separates out, which is collected and pressed between linen. The liquid is filtered through paper, all the alcohol removed by distillation from the water-bath, and the resin removed and dried at 100° Centigrade, until it ceases to lose weight. Dissolve it in double its weight of alcohol of 96 per cent, and perfume with lavender oil.

WALNUT STAIN FOR WOOD.—Water 1 qt., washing soda 1½ ozs., Vandyke brown 2½ ozs., bichromate of potash ¼ oz. Boil for ten minutes, and apply with a brush, either hot or cold.

WHITEWASH, To improve.—Add a strong solution of sulphate of magnesia.

WOOD, Red stain for.—A permanent and handsome reddish color may be given to cherry or pear tree wood by a coat of a strong solution of permanganate of potash, left on a longer or shorter time according to the shade required.

HINTS ABOUT DRAWING AND SKETCHING.

CAMERA LUCIDA, The.—This is probably the most reliable optical device employed for copying. The principle of its construction will be understood in the diagram marked 2 in the engraving. The glass is simply a four-sided prism, having one right angle, one of 135°, and two of 67½°. When disposed as represented, the rays from the object pass into it without any appreciable refraction, and are totally reflected from the lower inclined

FIG. 1.—THE CAMERA LUCIDA.

side, and again from the upper inclined side, emerging near the summit in a direction almost perpendicular to the top face, so that the eye sees on the paper placed beneath an image of the object. If the image be traced by the pencil, a very correct outline, not reversed, is obtained. The use of the device requires practice. The nearer the object copied is brought to the prism, the larger is its image, and *vice versa*.

A simple method of constructing the camera lucida is shown in

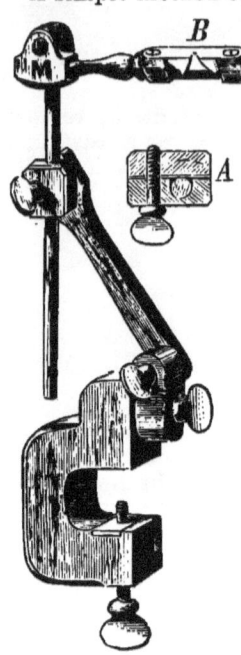

Fig. 2, and is the invention of Mr. H. E. Mead, artist of the SCIENTIFIC AMERICAN. The prism can be obtained at a small cost from any optician, and the rest of the apparatus any one can cut out of black walnut with a knife, and perhaps a gimlet. The thumbscrews used are of brass, of the kind employed for shutter-fastenings, and can be procured of any hardware dealer for a few cents each. B is the prism, and A is a section of one of the joints, showing how the apparatus may be easily adjusted. A movable rod, secured by a thumbscrew, regulates the height of the prism, and the single clamp shown secures it to the table. The cost of the whole is about one dollar.

DRAWING-BOARD, REFLECTING.—A flat board is provided, with two uprights, both of which, with the board, are grooved to hold a pane of glass in a perpendicular position. The drawing to be copied is secured to the board on the left of the glass, and the blank paper is fastened on the right. The artist now

stands to the left, as represented in the illustration, and looks down upon the glass at a very oblique angle. The original drawing is reflected from the polished surface of the pane to his eye, and at the same time he sees the white paper through the transparent glass, so that the lines of the model appear transferred, but reversed, upon the paper. These are followed with a pencil, and the outline is made.

DRAWING, Colors used in mechanical.—The annexed table shows different materials, and the colors used to denote them :

Cast-iron, . . .	Paine's gray and a little Indian ink.
" (another tint)	Ordinary neutral tint.
" "	Prussian blue and Indian ink.
Wrought-iron, . .	Prussian blue (or cobalt).
Steel, . . .	A purple made by mixing crimson lake and Prussian blue.
Gun-metal, . .	Gamboge or yellow cadmium.
Copper, . . .	Indian red mixed with a little lake.
Wood, . . .	Burnt umber.
Brick (red), . .	Indian red.
" (yellow), .	Indian yellow or cadmium, toned with white.
Stone color, . .	Chinese white and Indian ink, toned with yellow.
Water, . . .	Broken, irregular straight lines, with liquid copperas.

PAPER, TRANSFER.—A good transfer paper for copying monumental inscriptions and metallic patterns may be made by rubbing

A REFLECTING DRAWING-BOARD.

a mixture of black-lead and soap over the surface of common silver paper .

PAPER, TRACING, Temporarily transparent.—This is made by dissolving castor-oil in absolute alcohol, and applying the liquid to the paper with a sponge. The alcohol speedily evaporates, leaving the paper dry. After the tracing is made, the paper is immersed in absolute alcohol, which removes the oil, restoring the sheet to its original opacity.

PAPER, TRACING, that can be washed.—This is prepared by first saturating writing-paper with benzine, and then immediately coating it lightly with a varnish composed of boiled bleached linseed-oil 20 parts, lead shavings 1 part, oxide of zinc 5 parts, Venice turpentine ¼ part. Mix, boil for 8 hours, and, after cooling, add white gum-copal 5 parts, and gum-sandarac ¼ part.

PANTAGRAPH, The.—This consists of four rulers, jointed together at their intersections, and having, at two of the angles, supports terminating in round points or smoothly-running casters. At one of the other angles is a weight to which the apparatus is pivoted, and which holds it in place, and at the fourth corner is a tracing-point, shown in the hand of the operator. Directly across the frame thus made, and pivoted at its ends to the centres of two of the bars, is a fifth bar, through the middle of which passes a pencil. Along half the length of the two side-bars, and also of the central bar, are made perforations, so that the length of the rulers can be shortened as rendered necessary. The tracing point is moved over the outline to be followed, and its motion is communicated to the series of rulers, which, by a kind of parallel movement actuate the pencil to describe precisely the same

THE PANTAGRAPH.

line, equal in dimension to that of the copy, or enlarged or reduced. The scales of the two drawings are to each other as the distances of the pencil and of the tracing-point from the pivot, and these distances are adjusted by altering the position of the joints in the holes.

PENCILS, COPYING, To make.—Pencils are sold by stationers, the marks of which may be copied in the same manner as writing made by the pen with ordinary copying-ink. The method of preparing the leads is as follows : A thick paste is made of graphite, finely pulverized kaolin, and a very concentrated solution of aniline blue, soluble in water. The mixture is pressed into cylinders of suitable size and dried, when it is ready for use. Gum-arabic may be substituted for the kaolin.

SUN DRAWING.—Draw with a pencil on a piece of tracing-paper the desired design ; go over the lines with very black ink, turn the paper over, and follow the lines also with ink on the reverse side ; fasten the paper by the corners to a pane of clear glass. Make a solution of ½ oz. bichromate of potash in 2 ozs. hot water, strain when cold, and with this brush over the paper or silk on which the design is to be printed. Place the material thus prepared under the paper on the glass, and clamp all together. Expose the whole to bright sunlight, glass uppermost, then design, then bichromate paper ; in a few moments, the design will be printed deeply. Wash and soak for a short time in clean water (to fix), dry, and press with a warm flat-iron.

RULER, Perspective.—This is a simple arrangement for drawing lines in correct perspective. It consists in three arms of equal length pivoted together at one end by a screw-clamp. Two pins are inserted in the drawing-board, against which two arms of the ruler abut. The angle of these arms and the position of the pins are governed by the distance required for the vanishing-

THE PERSPECTIVE-RULER.

point, as the greater the angle, the further the same is removed, and *vice versa*. Once adjusted, the parts are clamped firmly together, and the lines ruled by the upper side of the arm which rests upon the paper. The arms at an angle are kept in contact with the pins, and the ruling arm is moved up or down the paper.

SKETCHING-FRAMES.—A square frame is hinged to the top of an ordinary drawing lap-board, so that it will stand in an upright position (as in Fig. 1). Across this is stretched a number of threads or wires at equal distances apart so as to divide the interior space into small squares. The paper on the board is similarly divided by light pencil-lines ruled over the surface. In making the sketch, the artist draws so much of the view as he sees through one of his squares in the frame, into the corresponding ruled square on the paper, and thus having a large number of straight lines to refer to is very readily enabled to locate the details of the picture. It is, however, necessary to use but one eye when looking at the landscape, and to keep this eye always at the same place, for which purpose an additional eye-piece may be added, simply consisting of a ring supported on a stand.

SKETCHING, OUT-DOOR, Simple apparatus for.—Provide a

small table with drawer; mount two grooved movable uprights at one end, with glass between the grooves; place an upright with a small eye-hole at the opposite end of the table, as shown in the engraving. Wash the glass with a thin solution of gum-

THE SKETCHING-FRAME.

arabic and rock candy (20 parts gum to 1 of candy). When the glass is dry, it is ready for use. Look through the small hole to get the object subtended by the glass, and with a soft crayon outline the subject on the prepared surface; remove the glass and lay it

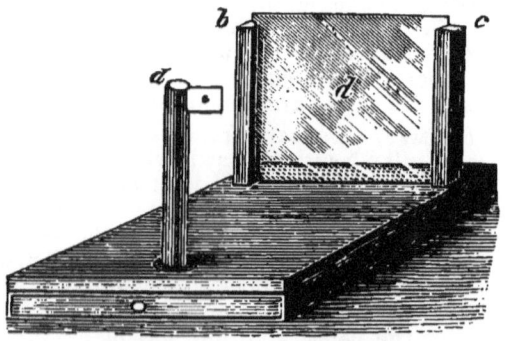

OUT-DOOR SKETCHING APPARATUS.

over your sketch. If you require the outline, you should have a second plate of glass, and trace over it the reverse way with charcoal, then lay your paper on, and a little gentle rubbing will transfer the outline.

TRACING-TABLE, TRANSPARENT.—This, as shown in the illustration, consists of a square-bottomed box, the tops of the sides of
which are inclined like those of a writing-desk.   The back is open,
and, as the apparatus rests on the table, abuts against a window.
The window-shade is drawn down to meet the upper part of the
device, so that the light enters through the back of the latter,

A TRANSPARENT TRACING-TABLE.

and the interior, being lined with white paper, is reflected up
through the inclined glass top ; the original drawing is laid upon
the glass, and a sheet of tissue-paper ruled off in squares is
placed above it.   Being brilliantly illuminated from below, the
drawing would readily show through, and might be copied, square
by square, as before described.

SIMPLE GALVANIC BATTERIES AND ELECTRO-PLATING RECIPES.

BATTERY CARBONS.—These can be readily cut with a hand-saw moistened in water.

BATTERY, DANIELL'S, Substitute for copper in.—Brighten sheets of ordinary sheet-tin and plunge into a very weak copper-plating solution, in connection with a galvanic battery of very low quantity. In 15 or 18 hours a tenacious film of copper will have been deposited upon the tin, and the plate can then be bent in shape suitable for the battery.

BATTERY, GALVANIC, A cheap.—Mr. W. M. Symons proposes a cheap but convenient galvanic battery : each of the zinc plates is 2 in. square, and covered with fustian or other fabric, outside which thick copper wire is wound to form the other plate ; the exciting liquid is weak chloride of zinc. Pairs of plates thus made can be arranged in series to form a battery to give out weak currents for a great length of time.

BATTERY, GALVANIC, A simple.—Take a glass tumbler, and place in the bottom a sheet of copper, having an insulated wire attached and extending out of the tumbler. Cover the copper with blue vitriol, and suspend a sheet of zinc near the top. Fill the tumbler with water. Connect the zinc and copper together for 48 hours, and the battery will be ready for use.

BATTERY, GALVANIC, Exciting liquid for.—Dissolve protosulphate of iron, 20 pts., by weight, in 36 pts. of water, stir in a dilution of sulphuric acid (equal parts of acid and water,) 7 pts., and add 1 part nitric acid similarly diluted. This liquid has great energy, and disengages no deleterious fumes.

BATTERY ZINCS, Amalgamation of.—The simplest and quickest method consists in immersing the zinc in a liquid composed of nitrate of mercury and hydrochloric acid. A few moments are sufficient for the complete amalgamation of the zinc, however soiled its surface may be. With a quart of this liquid, which costs less than 50 cents, 150 zincs can be amalgamated. The liquid should be prepared in this manner : Dissolve in warm water 200 grains of mercury in 1000 grains of aqua regia (nitric acid 1 part, hydrochloric acid 3 parts). When the mercury is dissolved, add 1000 grains of hydrochloric acid.

ELECTRO-MAGNETS, Softening.—Magnets or armatures for electro-motors may be softened as follows : Heat the iron to an even dull-red heat all over ; and if the surface of the iron has not been faced off in a machine, lightly file it to remove the scale, and then immerse it in common softsoap, allowing it to remain therein until it is quite cold. Then reheat the magnet to an even red heat whose redness is barely perceptible, and bury it in pulverized lime, wherein it must also remain until quite cold, when the metal will be found as soft as it is possible to make it, and the blade of an ordinary penknife will cut it. At the second heating, the iron will emit a light blue flame, showing the effect of the immersion in the softsoap. The capability of receiving strong magnetic power may be, by this process, very much increased.

ELECTRO-PLATED PAPER OR CLOTH.—Make a solution of nitrate of silver, and add ammonia until the precipitate formed at first is entirely dissolved. Place the paper or cloth for 1 or 2 hours in the liquid. After removing and drying, expose to a current of hydrogen gas, by which the silver is reduced to a metallic state, and the paper or fabric becomes so good a conductor of electricity, that it may be electro-plated with copper, silver, or gold in the usual manner.

ELECTRO-PLATING, Cleansing metals for.—Most articles are rapidly cleaned by chemical means. The first of these is the removal of grease by boiling in a solution of caustic soda, made by boiling 2 lbs. of common washing-soda and ½ lb. quicklime in a gallon of water ; after this they should be well brushed under water. The further processes will depend upon the nature of the objects.

1. Silver is washed in dilute nitric acid, then dipped for a moment in strong nitric acid, and well washed. Care must be taken that the water does not contain chlorine salts ; if the ordinary supply does so, the first rinsing after acids must be made in water prepared for the purpose by removing the chlorine by adding to it a few drops of nitrate of silver, and allowing the chloride to settle.

2. Copper, brass, and German silver are washed in a pickle of water 100 parts, oil of vitriol 100 parts, nitric acid (sp. gr. 1.3) 50 parts, hydrochloric acid 2 parts. . Spots of verdigris should be first removed by rubbing with a piece of wood dipped in hydrochloric acid ; they are then rinsed in water.

3. Britannia metal, pewter, tin, and lead can not be well cleaned in acids, but are to be well rubbed in a fresh solution of caustic soda, and passed at once, without washing, into the depositing solution, which must be alkaline.

4. Iron and steel are soaked in water containing 1 lb. oil of vitriol to the gallon, with a little nitric and hydrochloric acids added. Cast-iron requires a stronger solution, and careful rubbing with sand, etc., to remove scale and the carbon left by the acids. It is an advantage at times to connect them to a piece of zinc while cleaning. These metals should be cleaned just before placing in the depositing cell ; and if they are placed in an alkaline solution, they should be rinsed and dipped in a solution of caustic soda, to remove all trace of acids.

5. Zinc may be cleaned like iron, with a dip into stronger acids before the final washing.

6. Solder requires special care, as the acids used with the objects produce upon it an insoluble coating, and an obstinate resistance to deposit is set up at the edge of the solder. The same remark applies to soft-metal edgings and mounts. These should be rubbed with a strong caustic-soda solution, rinsed, and then treated as follows : Make a weak solution of nitrate of copper by dissolving copper in dilute nitric acid ; to a camel-hair or other soft brush, tie 3 or 4 fine iron wires to form part of the

brush ; dip this in the nitrate of copper, and draw over the solder, taking care that some of the iron wires touch it ; a thin adherent of copper will form, and upon this a good deposit will take place.

7. Old work for replating must have the silver and gold carefully removed ; if this is not done, there is apt to be a failure of contact at the edges of the old coatings, which causes blisters and stripping under the burnisher. The best mode of stripping is with the scratch-brush, etc., but chemical means may be used. Gold is dissolved by strong nitric acid, to which common salt is gradually added ; it may be collected afterward by drying and fusing with soda or potash. Silver is similarly dissolved by strong sulphuric acid and crystals of saltpetre, and recovered by diluting and precipitating with hydrochloric acid, then reducing the chloride either by fusion with carbonate of soda, or by acid and zinc cuttings. Copper can be removed from silver by boiling with dilute hydrochloric acid, and tin and lead by a hot solution of perchloride of iron.

INSULATORS, Rubber, Substitutes for.—Ivory and guaiacum wood, which are both relatively good conductors, become nearly non-conductive if stove-dried and saturated with certain oily and resinous liquids, which close up the pores of the bodies in question, and prevent moisture from penetrating within. Other kinds of wood can be modified in the same manner.

Sawdust of hard wood, agglutinated with blood and submitted to a considerable pressure, so as to mould it into a solid, tenacious body, is a good insulator for voltaic currents. After remaining six days in a damp cellar, it showed no galvanometric deviation.

IRON, Electro-plating with silver on.—*The direct way*: The article should first be rendered free from rust by rubbing with emery-cloth, or by dipping it into a pickle composed of sulphuric acid 2 ozs., hydrochloric acid 1 oz., water 1 gal. After the article has remained some time in this pickle, it should be taken out and the rust removed by a brush and wet sand. If the oxide can not be easily cleaned off, it must be returned to the pickle. As soon as the article is rendered bright, it is washed in a warm solution of soda, for the purpose of removing all grease. Lastly, it is well rinsed in hot water, and immediately placed in the plating solution, which should contain only about one fourth as much silver as that used for plating copper and brass articles. The battery power must also be weak. When the object receives a slight coating, the process may be carried on more rapidly by increasing the battery power, and by placing the article in a much stronger plating bath, using about 1 oz. silver in a gallon of solution. *The indirect method* consists in first coating the iron with copper, which insures success. Copper adheres firmly to iron, but silver does not ; hence copper acts the part of a go-between. After the article has been cleaned, as above described, it is coated with copper by placing it in a solution composed of carbonate of potassa 4 ozs., sulphate of copper 2 ozs., liquid ammonia about 2 oz., cyanide of potassium 6 ozs., water about 1 gal. The sulphate of copper may be dissolved in warm rain-water, and, when cold, the carbonate of potassa and ammonia added ; the precipitate when formed is redissolved. The cyanide of potassium should now be added, until the bluish color disappears. Should any precipitate be found in the bottom of the vessel, the clear solution may be poured off from it. The solution is worked cold, and with moderate battery power. Let the article remain in the bath until a thin film of copper is deposited, then remove quickly, rinse in hot water, and place in the silvering solution, where the process may go on as rapidly as if plating a copper article.

PEWTER, Electro-plating.—Take 1 ounce nitric acid, and drop pieces of copper in it until effervescence ceases; then add ½ ounce water, and the solution is ready for use. Place a few drops of the solution on the desired surface, and touch it with a piece of steel, and there will be a beautiful film of copper deposited. The application may be repeated if necessary, though once is generally sufficient. The article must now be washed and immediately be placed in the plating bath, when deposition will take place with perfect ease.

STEEL, Magnetization of.—If a properly-tempered steel needle be introduced into a magnetizing bobbin connected with a battery of constant current, battery and bobbin comprising the circuit, it acquires a total determined magnetism at the end of a period which appears not to exceed that of its introduction. On slowly withdrawing the needle, it is found to retain residual magnetism, which, together with the total magnetism, increases with each repeated introduction until a limit is reached. The needle may be magnetized in the bobbin by three other methods: 1. *Establishment* : Introduce the needle ; establish the current ; slowly withdraw the needle. 2. *Interruption* : With a closed circuit introduce the needle slowly ; break the current and withdraw the needle. 3. *Instantaneous charge* : Introduce the needle ; establish and break the current ; withdraw the needle. Repetitions of any of these three processes (all things being equal) insure an augmentation of the needle's magnetic moment. The last method is the best, but care must be taken to introduce the needle and current always in the same position, so as not to reverse the poles.

USEFUL CHEMICAL RECIPES FOR DETECTION OF ADULTERATIONS, FILTERING, INK-MAKING, ETC.

BEER, To prevent, from turning sour while on draft.—A slate cistern is made, having a wooden lid, fitting accurately, floating on·the surface of the liquid. The sides of the lid are beveled, so that a sharp edge is presented to the walls of the cistern, and along this edge a strip of india-rubber is fastened, which forms, with the bevel on the upper side, a V-shaped space, into which wet sand is packed in order to keep the rubber in close contact with the sides of the cistern, and so to exclude the air from the same. A hole is formed in the lid, having a stuffing-box, through which a pipe passes into the liquid, and the connection to the beer-engine is made in the usual way. The end of the pipe in the liquid is closed, but perforations are made in the sides about an inch therefrom : this prevents any sediment escaping with the fluid. Atmospheric pressure, acting on the lid, forces it to descend as the liquid is removed from under it, and thus a constant flow is obtained by means of the engine. By letting the cistern into the ground, the temperature of the liquid will remain nearly uniform the year round.

BEER, To clarify.—Take isinglass, finely shredded, 1 lb., sour beer, cider, or vinegar 3 or 4 pints ; macerate together till the isinglass swells, and add more of the sour liquid until a gallon has been used. Strain and further dilute. A pound of good isinglass should make 12 gallons finings, and 1½ pints finings is enough to clear a barrel of beer.

BISULPHIDE OF CARBON,To deodorize.—Distill with quicklime, the two substances having been in contact for 24 hours. The distillate is received in a flask partially filled with clean copper turnings. The lime remaining in the retort is strongly colored.

CAMPHOR, To powder.—Take camphor 5 ozs., alcohol 5 fl. drachms, glycerine 1 fl. drachm. Mix the glycerine with the alcohol, and triturate it with the camphor until reduced to a fine powder.

CANDLES, Paraffine.—To dye beautiful red, purple, or violet tints, use aniline colors.

CASKS, MOULDY, To disinfect.—Wash first for about 5 minutes with an alkaline solution of soda, and then soak for 1 or 2 days with a liquor acidulated with hydrochloric acid.

CHLOROFORM, Purifying decomposed.—Shake up the chloroform with a few fragments of caustic soda.

DYEING LEATHER YELLOW.—Picric acid dyes leather a good yellow, without any mordant ; it must be used in very dilute solution, and not warmer than 70° Fahr. Aniline blue modifies this color to a fine green.

DYES, TESTING, FOR ADULTERATION.—Red dyes must neither color soap-and-water nor lime-water, nor must they themselves become yellow or brown after boiling. This test shows the presence or absence of Brazil-wood, archil, safflower, sandalwood, and the aniline colors. Yellow dyes must stand being boiled with alcohol, water, and lime-water. The most stable yellow is madder-yellow ; the least stable are anatto and turmeric ; fustic is rather better. Blue dyes must not color alcohol reddish, nor must they decompose on boiling with hydrochloric acid. The best purple colors are composed of indigo and cochineal, or purpurine. The former test applies also to them. Orange dyes must color neither water nor alcohol on boiling ; green, neither alcohol nor hydrochloric acid. Brown dyes must not lose their color on standing with alcohol, or on boiling with water. If black colors have a basis of indigo, they turn greenish or blue on boiling with sodium carbonate ; if the dye be pure gall-nuts, it turns brown. If the material changes to red on boiling with hydrochloric acid, the coloring matter is logwood without a basis of indigo, and is not durable. If it changes to blue, indigo is present.

FILTERING, Hot.—The apparatus consists of a tube of soft sheet-lead, which can be wound around the funnel containing the filter in the form of a spiral. One end of the tube passes through a cork in the neck of a flask, in which water, or other liquid of higher boiling-point, is boiled ; the other end dips into a receiver, into which the condensed liquid flows.

FILTER, A simple.—The engraving represents a very good filtering apparatus. The best material for the box would be soapstone ; the next best material, iron. Mott's cast-iron tank-plates come of a convenient size—18 x 18 inches and 9 x 18 inches. These may be galvanized or coated with slate-paint.

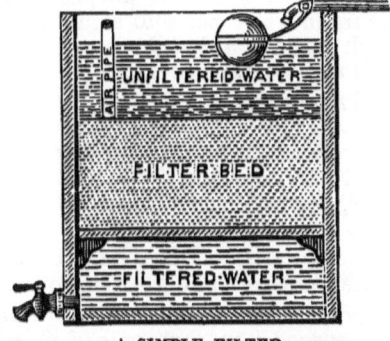

.A SIMPLE FILTER.

FREEZING-POWDERS. — (1) Four pounds sulphate of soda, 2½ pounds each of muriate of ammonia and nitrate of potash ; when about to use, add double the weight of all the ingredients in water. (2) Equal parts of nitrate of potash and muriate of ammonia ; when required for use, add more than double the weight of water. (3) Nitrate of ammonia and water in equal proportions. (4) Carbonate of soda and nitrate of ammonia equal parts, and 1 equivalent of water.

FILTERS, To make charcoal.—One method consists in pulverizing animal charcoal until reduced to an impalpable powder. This is mixed with a definite proportion of Norway tar and a compound of other combustible substances. The combined materials are then properly amalgamated with liquid pitch, and the whole kneaded up into a homogeneous plastic mass which admits of being moulded into slabs or blocks of any required dimensions and shape. These blocks, having been allowed to dry and harden, are subsequently carbonized by being subjected to a process of incineration by heat ; and in this manner all the combustible ingredients are burned out, leaving nothing behind but the animal charcoal in the form of a block of charcoal, permeated throughout by innumerable pores.

GASES, Drying.—Anhydrous phosphoric acid is the best substance known for this purpose.

GLYCERINE, Adulteration of, with sugar and dextrine, To detect. —To 5 drops of glycerine add 100 to 120 drops of water, 0.4 to 0.6 grain of ammonium molybdate, 1 drop pure nitric acid, and boil for a minute and a half. If any sugar or dextrine is present, the mixture will assume a deep-blue color.

GLYCERINE, Purification of.—To purify glycerine which has been for some time in use, add 10 lbs. iron-filings to every 100 lbs. of the impure liquid. Occasionally shake it and stir the iron. In the course of a few weeks, a black gelatinous mass will collect on the bottom of the vessel, and the supernatant liquid will become perfectly clear, and can be evaporated to remove any excess of water that may have been added to it.

GLYCERINE, Testing.—When treated slowly with sulphuric acid, it should not turn brown ; with nitric acid and nitrate of silver, it should not become cloudy ; and when rubbed between the fingers it should not emit a fatty smell.

HIDES, To preserve.—Carbolic acid is used in South-America and Australia for this purpose. The immersion of hides for 24 hours in a two per cent solution of carbolic acid, and subsequently drying them, has been successfully substituted for the more tedious and expensive process of salting.   .

HYDROCARBONS, Classification of.—The classification usually adopted by distillers is as follows : All above 88° of Baumé's hydrometer is called chymogene, from 88° to 70° gasoline, from 70° to 60° naphtha, from 60° to 50° benzine, from 50° to 35° kerosene, from 35° to 28° lubricating-oil.

INK, BLUE.—Prussian-blue 6 parts, oxalic acid 1. Mix with water to a smooth paste. Dilute with rain-water, and add a little gum-arabic to prevent spreading.

INK, COPYING, used without a press.—Coarsely-broken extract of logwood 1 oz., carbonate of soda (crystallized) 1 drachm : heat in a porcelain capsule with 8 ozs. distilled water until the solution is of a deep-red color. Remove from fire and stir in glycerine 1 oz., neutral chromate of potash dissolved in a little water 15 grains, and a mucilaginous solution of 2 drachms finely-pulverized gum-arabic. Keeps well, never requires a press for copying, and does not attack steel pens. The impression is taken on thin moistened copying-paper, at the back of which is placed a sheet of writing-paper.

INK FROM ELDERBERRIES.—Bruise the berries, place them in an earthen vessel, and keep in a warm place for 3 days. Press out and filter. Add to 12½ ozs. of this filtered juice 1 oz. sulphate of iron and the same quantity of pyroligneous acid. This ink is violet at first, and afterward becomes black.

INK, INDELIBLE, for marking linen.—(1) Bichloride of copper, 8¼ grains, dissolved in distilled water, 30 grains ; then add common salt, 10 grains, and liquid ammonia, 9½ grains. A solution of 30 grains hydrochlorate of aniline in 20 grains distilled water is then added to 20 grains solution of gum-arabic, containing 2 pts. water, 1 pt. gum-arabic, and 10 grains glycerine. Four parts of the aniline solution thus prepared are mixed with 1 part of the copper solution. This ink can be used with a steel pen. It is green at first, but becomes black in a few days or by application of hot iron. It is absolutely indelible, and the finest devices can be written with it. It is better to mix the two solutions only just before using. (2) For very fine linen, take a solution of nitrate of silver, 4 pts., in distilled water, 24 pts. Add liquid ammonia until the precipitate formed is dissolved. Then a little sap-green and indigo are ground together and mixed with a solution of gum-arabic, 4 pts., and this is mixed with the nitrate of silver solution. The whole is then diluted until it occupies 32 parts. It turns black as No: 1 does.

INK, To restore dim.—Cover the letters with solution of ferrocyanide of potassium, with the addition of diluted mineral acid (muriatic) ; upon the application of which the letters will change to a deep-blue color. To prevent the color from spreading, the ferrocyanide should be put on first, and the dilute acid added upon it.

INKS, ANILINE.—Violet ink is obtained by dissolving one part of aniline violet-blue in 300 parts of water. This ink is quite limpid, dries quickly, and gives a remarkably dark color. It is necessary that new pens should be employed in using it, as the smallest quantity of ordinary ink mixed with it causes its alteration. Blue ink is made by dissolving 1 part of soluble Paris-blue in 250 parts of boiling water ; red ink, by dissolving 1 part of soluble fuchsin in 200 parts boiling water. While ordinary inks are decomposed by numerous substances, and notably by hydrochloric acid, aniline inks are completely ineffaceable from the paper on which they are used. They resist the action of acids, and even of chlorine.

INKS, Sympathetic.— *Yellows* (1) : Sulphate of copper and sal-ammoniac, equal parts, dissolved in water ; (2) onion-juice ; both visible on heating   *Black* (1) : Weak infusion of galls. This is turned black by weak solution of protosulphate of iron. (2) Weak solution of protosulphate of iron. Turns blue when moistened by weak solution of prussiate of potash, and black by infusion of galls. *Brown :* Very weak solutions of nitric, sulphuric, muriatic acids, common salt, or nitrate of potash. Visible on heating. *Green :* Solution of nitro-muriate of cobalt. Brought out by heat ; fades when cool. *Rose-red :* Acetate of cobalt solution, with a small quantity of nitrate of potash. Acts as preceding. Solutions of nitrate of silver and terchloride of gold become permanently dark when exposed to sunlight.

INK, White, for colored paper.—1 part muriatic acid, and 20 parts starch-water. Very dilute oxalic acid may also be used. Write with a steel pen.

IVORY AND BONES, Bleaching.—Spirit of turpentine is very efficacious in removing the disagreeable odor and fatty emanations of bones or ivory, while it leaves them beautifully bleached. The articles should be exposed in the fluid for 3 or 4 days in the sun, or a little longer if in the shade. They should rest upon strips of zinc, so as to be a fraction of an inch above the bottom of the glass vessel employed. The turpentine acts as an oxidizing agent, and the product of the combustion is an acid liquor which sinks to the bottom, and strongly attacks the bones if they be allowed to touch it. The action of the turpentine is not confined to bones and ivory, but extends to wood of various varieties, especially beech, maple, elm, and cork.

IVORY, Imitation.—To liquid chloride of zinc of 50° to 60° Baumé, add 3 per cent of sal-ammoniac ; then add zinc-white until the mass is of proper consistence. This cement may be run into moulds, and when hard becomes as firm as marble.

LIGHT, brilliant white, To make a.—Fill a small vessel of earthenware or metal with perfectly dry salpetre or nitre, press down a cavity into its surface, and in this cavity place a piece of phosphorus; ignite this, and the heat given off melts a sufficient quantity of the nitre to evolve oxygen enough to combine with the phosphorus, and the effect is to produce the most magnificent white light which chemistry can afford.

MICA.—The best comes from the Eastern States. New-York mica is good. Canada mica is of several different shades, from light brown to intensely black.

OIL, COTTON-SEED, Refining.—One hundred gallons of the crude oil are placed in a tank, and 3 gallons of caustic potash-lye, of 45° Baumé, are gradually added and well stirred for several hours; or the same quantity of oil is treated with about 6 gallons of soda-lye of 25° or 30° Baumé, and heated for an hour or more to about 200° or 240° Fahr., under perpetual stirring, and left to settle. The clear yellow oil is then separated from the brown soap stock, and this dark soap sediment is placed into bags, where the remainder of the oil will drain off; and the sediment has a marketable value of 3 or 4 cents a pound for soap-makers. The potash-lye has to be made in iron pots, but the oil and lye may be mixed in wooden tanks.

OILS FROM PLANTS, odoriferous, Extraction of.—This can be done by glycerine. The flowers are introduced into the liquid and left for 3 weeks. The glycerine is then drained off, and may be dissolved in all proportions in alcohol or water to make perfumed liquids or washes.

OILS, LUBRICATING, Testing for acids in.—Dissolve a crystallized piece of carbonate of soda about as large as a walnut in an equal bulk of water, and place the solution in a flask with some of the oil. If, on settling after thorough agitation, a large quantity of precipitate forms, the oil should be rejected as impure. As oils are often clarified and bleached with acids, which injure the metals on which they are used, this is an important point to know.

OIL, SPERM, To prevent gumming.—It may be purified by agitating 100 parts oil with 4 parts chloride of lime and 12 water; a small quantity of decoction of oak-bark is afterward added to remove all traces of gelatinous matter which it retains, and the mixture is left to settle. The clear oil is afterward agitated with a small portion of sulphuric acid, again clarified by subsidence, and washed to remove adhering sulphuric acid. The addition of mineral oils, as heavy kerosene, has also the tendency to prevent gumming, or at least greatly to diminish it.

OILS, VOLATILE, Explosion of.—A mixture of 2 parts of perfectly dry permanganate of potassium with 2 or 3 parts of concentrated sulphuric acid is a most powerful oxidizing agent, owing to the separation of permanganic acid and its immediate decomposition with the liberation of oxygen. Volatile oils are violently affected by this mixture, if about 10 drops are placed in a little dish and then touched with a stout glass rod previously dipped into the mixture. The following produce explosions, often most violently : oils of thyme, mace, turpentine (rectified), spike, cinnamon, origanum, rue, cubebs, and lemon. The following oils are simply inflamed, particularly if poured upon·blotting-paper and touched with the mixture, though under certain still unknown circumstances explosion may occur : oils of rosemary, lavender, cloves, rose, geranium, gaultheria, caraway, cajeput, bitter-almond, and rectified petroleum. The following substances are ignited without explosion : alcohol, ether, wood-spirit, benzole, chlorelayl, sulphide of carbon, and cotton. Gun-cotton and gunpowder are not ignited.

PETROLEUM, Test for illuminating.—Fill a tumbler full of water at 110° Fahr. Stir in a tablespoonful of the oil to be tested, and leave until the oil reaches about the same temperature. Pass a lighted match over the oil as it floats on the surface. If the oil does not ignite, it can be safely used ; if it does, discard it, however cheap the price may be. Improved test proposed by Dr. Van der Weyde : Fill a narrow test-tube with the petroleum to be tested, close it with the finger, invert it, and plunge entirely in water of some 140° Fahr. ; wait until the temperature has descended to 110° ; if then any gas-bubbles are seen in the closed upper part of the test-tube the oil contains dangerous inflammable vapors. As all vapors of petroleum are inflammable, it is not necessary to ignite them ; the demonstration of their presence in this way is sufficient to condemn such oil.

RAW HIDE, To dissolve.—This can be done completely in water heated under pressure.

RESINS, Solubility of.—Copal, amber, dammar, colophony, lac (or shellac), elemi, sandarac, mastic, and carnauba wax (a resin) have been experimented upon. Amber, shellac, elemi, sandarac, and mastic swell up and increase in bulk when heated ; the others fuse quietly. Carnauba wax melts in boiling water, colophony becomes pasty therein, while dammar, shellac, elemi, and mastic agglutinate. Copal, amber, and sandarac do not change in water.

Alcohol does not dissolve amber or dammar ; it agglutinates copal, and partly dissolves elemi and carnauba wax ; while colophony, shellac, sandarac, and mastic are readily soluble therein.

Ether does not dissolve amber and shellac ; it makes copals swell, and partly but slowly dissolves carnauba wax ; it readily dissolves dammar, colophony, elemi, sandarac, and mastic.

˜Acetic acid does not dissolve amber and shellac ; it causes copal to swell ; it somewhat acts upon carnauba wax, but not at all upon any other of the resins above named. A hot solution of caustic soda, of sp. gr. 1.074, readily dissolves shellac, with difficulty colophony, and has no action upon the rest. In sulphide of carbon, amber and shellac are insoluble ; copal swells therein ; elemi, sandarac, mastic, and carnauba wax are with difficulty dissolved, while dammar and colophony are readily so. Oil of turpentine has no action upon amber or shellac ; it causes copal to swell, and readily dissolves dammar, colophony, elemi, sandarac, carnauba, and very readily mastic. Sulphuric acid does not dissolve carnauba wax ; it dissolves and colors all other resins brown, except dammar, which becomes bright red. Nitric acid does not act upon the resins, but covers carnauba wax straw-yellow, elemi dirty yellow, and mastic and sandarac bright brown. Ammonia does not dissolve some of these resins, but causes copal, sandarac, and mastic first to swell, afterward dissolving them ; colophony is easily soluble therein.

RUBBER, Solvents for—These are ether (free from alcohol), chloroform, bisulphide of carbon, coal naphtha, and rectified oil of turpentine. By long boiling in water, rubber softens, swells, and becomes more soluble in its peculiar menstrua ; but when exposed to the air, it speedily resumes its pristine consistence and volume. Oil of turpentine dissolves caoutchouc only when the oil is very pure and with the application of heat ; the ordinary oil of turpentine of commerce causes india-rubber to swell rather than to become dissolved. In order to prevent the viscosity of the india-rubber when evaporated from its solution, one part of caoutchouc is worked up with two parts of turpentine into a thin paste, to which is added $\frac{1}{4}$ part of a hot concentrated solution of sulphuret of potassium in water ; the yellow liquid formed leaves the caoutchouc perfectly elastic and without any viscosity. The solutions of caoutchouc in coal-tar, naphtha, and benzoline are most suited to unite pieces of caoutchouc, but the odor of the solvents is perceptible for a long time. Sulphide of carbon is the best solvent for caoutchouc. This solution, owing to the volatility of the menstrum, soon dries, leaving the latter in its natural state. When alcohol is mixed with sulphide of carbon, the latter does not any longer dissolve the caoutchouc, but simply softens it and renders it capable of being more readily vulcanized. Alcohol also precipitates solutions of caoutchouc. When caoutchouc is treated with hot naphtha distilled from native petroleum or coal-tar, it swells to 30 times its former bulk ; and if then triturated with a pestle and pressed through a sieve, it affords a homogeneous varnish, the same that is used in preparing the patent water-proof cloth of Macintosh. Caoutchouc dissolves in the fixed oils, such as linseed-oil, but the varnish has not the property of becoming concrete on exposure to the air. Caoutchouc melts at a heat of about 256° or 260° after it has been melted ; it does not solidify on cooling, but forms a sticky mass which does not become solid even when exposed to the air for months. Owing to this property, it furnishes a valuable material for the lubrication of stop-cocks and joints intended to remain air-tight and yet be movable.

RUBBER, To cut.—Dip the knife-blade in a solution of caustic potash.

VINEGAR, Making, from alcohol (Artus's process).—Dissolve $\frac{1}{2}$ oz. dry bichloride of platinum in 5 lbs. of alcohol. With this moisten 3 lbs. of charcoal broken to the size of a hazel-nut. Heat the charcoal in a covered crucible, and place it in the bottom of a vinegar-vat. This causes the rapid oxidation of the alcohol. Reheat the charcoal once in 5 weeks.

THE FARM.

FARM BUILDINGS.

BEAMS, Fastening in walls.—The usual custom of building the ends of floor-timbers into brick and stone walls is apt, in case of fire, to throw over the walls; and resting the timbers on corbels interferes with the cornice-line below. By cutting the ends of the timbers on a bevel and laying in the wall, as in the annex-

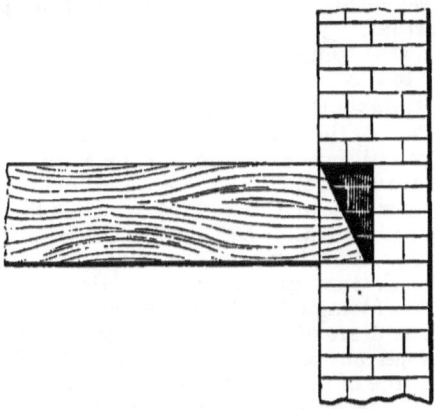

SETTING BEAMS IN WALLS.

ed diagram, the cornice-line will not be broken; and, in case of fire, the timbers will fall with little injury to the wall.

BLASTING.—In small blasts, 1 lb. of powder will loosen about 4½ tons of rock. In large blasts, 1 lb. of powder will loosen 2½ tons. 50 or 60 lbs. of powder inclosed in a bag and hung against a barrier will demolish any ordinary structure. One man can bore with a bit 1 in. in diameter, from 50 to 60 in. per day of 10 hours in granite, or 300 to 400 in. per day in limestone. Two strikers and a holder can bore with a bit 2 in. in diameter 10 ft. per day in rock of medium hardness.

BRICKS FROM GAS-COAL ASHES.—These are of remarkable lightness, porosity, and dryness. The ashes, after being taken from the retorts, are spread on the surface of a clean floor; they are then finely pulverized, and 10 per cent of slaked lime, together with a small proportion of water, is intimately stirred and incorporated with them. After a rest of 24 hours, the mixture is made into bricks by the ordinary process. The bricks are immediately transferred to the drying sheds, where a few days' exposure renders them fit for use.

BRICKWORK, Preserving.—To exclude dampness, use the following : ¾ lb. mottled soap is dissolved in 1 gall. boiling water, and the hot solution spread steadily with a flat brush over the outer surface of the brickwork, care being taken that it does not lather ; this is allowed to dry for twenty-four hours, when a solution, formed of ¼ lb. alum dissolved in 2 galls. water, is applied in a similar manner over the coating of soap. The soap and alum form an insoluble varnish, which the rain is unable to penetrate, and this cause of dampness is thus said to be effectually removed. The operation should be performed in dry, settled weather.

Another method is to use 8 parts linseed oil and 1 part sulphur, heated together to 278°, in an iron vessel.

CHIMNEYS, SMOKY, Causes of.—Want of sufficient height in the flue. The outlet of the chimney being placed in an exposed and cold situation, while the air with which the fire is supplied is drawn from a warmer and more sheltered region. Excessive width in the flue, by which a large volume of cold air is drawn in and allowed to lower the temperature of the ascending column. Low temperature of the interior of the flue, in comparison with that of the external air. Humidity of the air. Too accurate fitting of the windows and doors, and joints in the flooring. The draft of one fire injuring that of others in the same house. A current caused by the heat of the fire circulating in the room. A flue of insufficient size. A foul flue. Displacement of masonry, or accumulation of mortar within the flue. The sudden obstruction of the draft, by gusts of wind entering the chimney-top. Increase of density of the air at the chimney-top, due to the effect of wind in chimneys rising from the eaves of roofs. Drafts within the room which throw the smoke out of the influence of the ascending chimney current.

CHIMNEYS, SMOKY, Preventing effects of.—A screen or blower of wire gauze, from 36 to 40 wires to the inch, placed in front of range or stove fires, will prevent, it is said, smoke coming into the room when the chimney fails to draw well.

CISTERN, Building a.—One thing is essential, and is very generally neglected. It is to have the water as it comes into the cistern conducted to the bottom. In this way, the water is entirely changed when it rains. When the fresh water simply pours in at the top, it immediately runs off, and all the mass of stagnant water remains undisturbed, and soon becomes impure.

CISTERN FILTER.—A wall of soft burned bricks is well adapted for this purpose, when built up within the cistern.

CONCRETE FOUNDATIONS, To build.—The concrete is composed of lime, sand, water, gravel, and round or broken stones. A trench of boards is first made, of the width of the desired foundation. Fill the trench with the concrete to the depth of a foot or two, and let it stand until sufficiently hard ; then add another foot of concrete, and so go on, adding concrete and raising trench-boards as the wall rises.

CONCRETE PAVEMENTS.—The cheapest material for mixing with gravel is coal-tar from gas-works.

GREENHOUSE, To build a cheap.—Mr. Peter Henderson says the ordinary span-roof is best. The walls are 4 ft. high, formed of locust or cedar posts. To the outside of these are nailed boards—rough hemlock will do, if appearances are not considered. To the boards is tacked the ordinary tarred paper used by roofers. Against the paper is again nailed the outer or weather boarding. This makes really a better wall for greenhouse purposes than an 8-inch one of brick, as we find that the extremes of temperature of the greenhouse—inside at 50°, and perhaps 10° below zero outside—very soon destroy an 8-inch solid brick wall, particularly if exposed to the north or west. A wall of wood constructed as above will last for twenty years, and be as good a protection as one of 8-inch brick. The roof is formed by the ordinary sashes, 6 ft. in length by 3 ft. in width, which can be bought ready made. Heat with a flue not more than 60 and not less than 30 ft. in length; if more, the flue would not heat it enough, and if less it would be likely to get too much heat. About 50 ft. by 11 is, we think, the best size of a greenhouse to heat with a flue. The flue should run all around the house—that is, it should start along under one bench, cross the end, and return under the other bench to the end where it begins, making the length of flue in a greenhouse of 50 feet about 110 feet long. It should have a "rise" in this length from the furnace of at least 18 in., to secure a free draught. For the first 25 ft. of flue nearest the furnace it should be built of brick, forming an air-space inside of about 7 x 7 in. From this point (25 ft. from the fire) the flue should be formed of the ordinary drain-pipe cement or terra-cotta. The former is to be preferred, and that of 7 or 8 in. diameter is best. The cost of a greenhouse thus built in the vicinity of New-York, is about $6 per running foot—that is, one 50 ft. long by 11 ft. wide costs $300.

ICE-HOUSE, To build an.—A house 12 ft. square by 8 or 9 ft. high is large enough for a good-sized family. It may be a frame building, entirely above the surface of the ground, and better if supported on posts, elevated a few inches, to be certain of good drainage. Build of joists, 2 x 3 in., with an outer boarding, having inside another series of uprights, also boarded, from 6 to 10 in. removed from the outer shell, with a solid floor of plank, the space between the two walls filled with tan, sawdust, straw or chaff, and a roof of good pitch. A drain for water should be made from the floor, and the space above the uprights, between a loose flooring and the pitch of the roof, filled with straw, hay, or some other dry, porous material. On the roof should be a ventilator, the top defended from rain or snow. The ice should be packed in one solid mass, the sides not reaching the inner walls, but allowing a space of from 6 to 12 in. all around. The top of the ice should be covered with straw, and the door should be like the sides of the building, or double doors should be made, one in the outer and the other in the inner wall. Plant morning-glories or any other climbing plant about the building, and train them up over the roof, so that their foliage will serve as a protection against the sun.

LIGHTNING-RODS, Valuable hints concerning.—Feather-beds
are not a protection from lightning. The human body is a
better conductor of electricity than feather-beds or other objects
ordinarily contained in the apartments of dwellings, and therefore, when the lightning enters an apartment, the human body is
likely to form one in a chain of inductions, determining the path
of an electrical discharge, unless better conductors are in its vicinity to divert this action.

The only place of absolute security in a thunder-storm is an iron
building ; or next in safety is a building properly protected by
lightning-rods.

A copper rod of one inch in diameter, or an equal quantity of
copper under any other form, will resist the effect of any discharge
of lightning hitherto experienced. The copper rod is therefore
the safest and best material that can be used, but it is expensive.
Iron rods of one inch in diameter are very commonly used, and,
if pointed with solid copper and properly put up, are efficacious
in the great majority of cases. The particular form of the rod
makes no difference. It may be round or square, twisted or hollow, composed of one solid piece or made of wires twisted together. It is the quantity of metal contained in the cross-section
of the rod that is of value, not the form.

Lightning-rods are provided with sharp points to allow the ac-
cumulated negative fluid to pass off readily into the air and neu-
tralize the positive fluid of the thunder-cloud.

The object being to make so good a passage for the lightning
to the ground as to remove all danger of its leaping to some conductor in the house, the greatest care must be taken not to have
any break in the conductivity. As it is inconvenient to manufacture or transport the rods in one piece, the different parts must be
in intimate connection when they are put up ; it is best to have
them soldered, and the joints protected from the air and moisture.

The point of the rod should be extended a little above the chimney or highest part of the building, and should be fastened in
contact with the building by staples or cleats. Glass insulators
should not be employed. It makes no difference in conductivity
whether the rod is painted or not painted.

No building can be said to be properly rodded or protected
against lightning, unless the lower part of the rod or terminal
under the ground is made quite extensive. The extremity of the
rod should connect with masses of good conducting materials,
such as old iron, or iron ore, or coke, or charcoal, laid in trenches,
or the rod itself should be elongated, sunk deep in the ground,
and carried a considerable distance from the building, and put in
connection with water, or moist earth if possible. The golden
rule for safety is : "Provide the largest possible area of conducting surface for the terminal of the rod."

A lightning-rod which is not properly connected with the earth
is quite dangerous. The very common method of merely sticking the lower end of the rod down into the dry earth near the
surface of the ground is bad, and endangers the building, because
dry earth is such a poor conductor, and the amount of rod surface
in contact with the earth is so small. Under such conditions, a
portion of the electric current will be likely to find an easier path

to the earth through the building than through the rod; and a part of the electricity will therefore leave the rod, strike into the building, and down in various directions into the earth, making havoc as it goes. As a measure of prudence, house-owners should look to the terminals of their lightning-rods, and place there a considerable amount of the conducting materials above named.

It was supposed to have been established by Charles and Gay Lussac that a lightning-rod protected an area whose radius was double the height of the rod extending above the building; but this rule is no longer reliable, by reason of the extensive use of metals in the shape of pipes, etc., in the construction of the buildings of our day.

When electricity finds several paths to the ground, it will prefer the best, it is true; but some portion will also pass along the poorer conductors If, therefore, any metallic substances lie within the area supposed to be protected, they are in danger of being struck. This is especially true where the lightning has a chance to jump to the gas and water pipes of a building. It is a good plan to connect these pipes with the lightning-rod; if the rod is struck, the electricity will then have an excellent path into the ground, and will be rapidly diffused over the vast underground network of pipes. The danger to the inmates of the house of being struck from these pipes is less than that of receiving a shock from the powerful induced currents liable to be developed in them, if unconnected, during a thunder-storm.

The more rods on a building the better, especially if all are connected with each other near their upper ends.

Finally, in the way of general advice, we would say: Connect all your lightning-rods together, and also to your iron tank, and water, gas, or other pipes, not by separate connections, but so that there is some connection between all, which connection should be as high up as possible. If you have a metal roof, connect all rods with it. If the roof is not of metal, then connect your rods together by means of a good-sized conductor running along the ridge of the roof. Bear in mind that, to carry off the heaviest lightning-flash known, a copper rod one inch in diameter is not considered too large; and though of course such flashes are of very rare occurrence, they may come. Hence the great value of uniting your different rods high up.

MORTAR, Good weatherproof.—3 bushels clean sand, mingled with ½ bushel good lime and ½ bushel cement, makes an excellent mortar which is not liable to be dislodged by storms.

MORTAR, To make.—The lime ought to be pure, completely free from carbonic acid, and in the state of a very fine powder; the sand should be free from clay, partly in the state of fine sand and partly gravel; the water should be pure, and, if previously saturated with lime, so much the better. The best proportions are 3 parts fine sand, 4 parts coarse sand, 1 part quicklime recently slaked, and as little water as possible. There should always be enough water added at first; if water is added after slaking has begun, it will be chilled and the mortar lumpy. The addition of burnt bones improves mortar by giving it tenacity, and renders it less apt to crack in drying.

OAK TIMBER, Seasoning.—Oak loses about $\frac{1}{8}$ its weight in seasoning, and about $\frac{1}{4}$ its weight in becoming perfectly dry.

PAVEMENT, Farmyard.—Make a concrete of gravel or sand and Portland cement; or easier, of gravel, sand, coal-ashes, and coal-tar. Dig away the earth for 5 in., lay a bottom of pebbles as large as goose-eggs, ramming well down. Sweep off clean, and pay the surface with hot coal-tar, thinly; put on a coat of smaller gravel previously dipped in hot tar, drained, and rolled in coal-ashes with an intermixture of gravel. Roll it down as compactly as possible. Let the roller run slow, and let a boy follow it with a hoe to scrape all adherent gravel. Next, put on a coat of fine gravel or sand, coal-tar and some coal-ashes, to complete the surface. Roll again. This will take some weeks to harden, but will shed water, and eventually form a very firm surface. Do not use too much tar, but only enough to make the ingredients cohere under pressure.

RAT-PROOF BUILDINGS.—The plan adopted in England is to have slate floors, sawed and planed to uniform sizes and thickness. The walls are also covered with sawed or planed slates, well jointed and secured to the wall or studding with screws, which makes each room as secure against rats as an iron or stone box would be. The slate used for the floor is from 1 to 2 in. thick, and that for the walls $\frac{1}{2}$ in. thick.

RAT-PROOF FRAME BUILDINGS.—Nail strips of board to the sill between each flooring joist, on the inside, reaching to the under side of the flooring planks or boards, and thereby covering the shelf formed by the sill between the joists. The idea is to allow the rats no place to stand upon while they are cutting through the floor.

ROADS, CORDUROY, To build.—First lay all small poles or brush transversely and across the road. Next take long trees—the smallest ends being at least 10 inches diameter—and place them longitudinally across the poles, in two rows, 8 feet apart from centre to centre, making the ends at the junction of each piece lap each other at least 3 feet, breaking joint on either side, and placing upon these ends large logs of sufficient length to extend across the road, and 2 feet on each side of these stringers. Cover the stringers with transverse logs, 12 feet long from scarf to scarf, and at least 10 inches in diameter at the smallest end, fitted close together on the straight portions; the logs alternated with a large and small end, and on the outer side of curves, all the large ends, which will assist in the curvature of the road and the gravity of the vehicles. Next, adze off the centre ridges of these logs to a face of about 5 inches, for a width of 9 feet in the centre of the roadway, and cover this 9 feet with gravel, to fill in between the logs, and give a smooth surface. A good plan is to lay on the top of the road thus formed poles of 5 or 6 inches in diameter, spiked down on each side of the track, every 10 feet, with oak pins, to prevent, in frosty weather, the lateral sliding of wagons.

ROOFING, Pasteboard and asphalt.—This material is most suitable for flat roofs, having a fall of $1\frac{1}{2}$ inches to $4\frac{1}{2}$ inches per running foot. It may, however, also be used for roofs having a greater fall, the expense being in this case somewhat larger than for flat roofs, as the laying on is more difficult. Cover the roof first with dry boards, ¾ inch to 1 inch thick, and not above 6 inches broad; if more than the last-named width, or if not sufficiently dry, the boards ought to be split once before being laid on, in order to keep them from warping, and every board should be fastened with three nails at least on each of the rafters. The boards do not require to be rabbeted; only those ends of the boards which form the eaves, by extending beyond the wall need to be joined in the said manner. In case of boards ¾ inch thick being applied, the rafters should not be more than 2 feet from each other, as the boards may be too elastic and not strong enough to support the weight of the workmen, while the roof will not be perfectly substantial.

The roofing may be done either from gable to gable, or from the eaves to the foot-ridge, the first roll being laid with a bend of 1 inch beyond the roof, and fastened with the flat-headed iron wire nails supplied for that purpose. The second roll is laid 1½ inches over the first, and so on till the roof is covered. The joints and heads of the nails are then coated with asphalt mastic, and the seams thus coated are strewed with dry sand. The whole roof is then coated with the mastic, and covered with sand. This coating, which is only to be effected in dry weather, renders the roof perfectly water-tight, and it can then, if desired, be painted or whitewashed. A hundredweight of mastic covers a surface of 65 square yards. This process is in use in Copenhagen, Denmark, and the roof weighs about ⅓ the weight of a tiled roof, and is substantial, resisting alike the influence of water, fire, heat, and cold.

ROOFING, Portland cement and tar.—The inclination of the framework of the roof (which must have an even surface) should be at the rate of from ⅓ to ¾ inch per foot. The rafters or joists should not be more than 2 feet 3 inches apart, so as to give sufficient strength. As the rafters rest on the side walls, a comparatively small quantity of timber is required. Boards of 1 inch or 1¼ inch thick are fastened or nailed on the rafters, and should be dovetailed. These are then covered with a layer of sand ¼ or ½ inch thick, in order to produce an even surface. Strong brown paper, in continuous rolls, and as broad as possible, is then laid upon it, so that each length overlaps the other by about 4 inches. When the whole or a large part has thus been covered with paper, the mixture is put into a caldron, in the proportion of tar 100 pounds to Portland cement 180 pounds. Care must be taken to heat the tar gently, and to mix the cement with it gradually, in order to prevent its boiling over. This mixture of tar and cement must then be laid on as hot as possible on the paper with a tar-brush. The next layer of paper is then laid upon it, and smoothed with a light wooden roller. In this way the whole roof must be covered. In order to break the joints of the paper, begin the second layer with half the breadth, and proceed as before. The third and fourth layers are laid in like manner, with alternate layers of cement and brown paper. The last layer must be carefully covered with cement, and then strewn with sifted ashes to the thickness of $\frac{1}{4}$ inch. Next to the gutter is a board covered with zinc, and projecting about 2 inches. It should be laid on after the second layer has been completed, so as to be covered by the third and fourth. If there are any chimneys projecting through the roof, they should be surrounded with zinc immediately after the first layer has been finished.

ROOFING ZINC.—Permit perfect freedom to the sheets. Confine them nowhere, and separate lengths of guttering, and any other portions of a roof requiring to be made in long pieces, as much as possible. Eaves-gutters should be made in short lengths, bent in the direction of the way in which the sheet has been rolled and soldered, the solder being put between the sheets, and one sheet lapping over the other. Nor must they be screwed to the rafters, as this is a practice which occasions a constant failure in the joints of the iron eaves-gutters. Wherever a down-pipe comes, there should be a stopped end in the gutter; and the gutter should not be continued longer than possible in one place. Where it is laid behind a parapet, a separate piece of flashing will disconnect it wholly from the sheeting on the roof. For guttering, the gauge used should be increased in proportion to length; there should be a proper substance in all cases. Oak boarding will spoil the zinc, and the fir boarding should be dry—the boards with an aperture of about $\frac{1}{4}$ inch between each. If they are damp, as much oxidation will take place on the under side of the zinc as on top of it. From experiment, it appears that the oxidation proceeds for about four years, gradually diminishing after the first three months, when it hardens into a protecting coat of a dark gray color, preserving the metal beneath from further deterioration. A sheet of zinc exposed to the atmosphere for a series of years loses little or nothing of its weight or thickness, and its surface remains hard and polished as enamel.

SHINGLES, To prevent decay of.—Put into a large tub 1 barrel of wood-ashes lye, 5 pounds white vitriol, 5 pounds alum, and as much salt as will dissolve in the mixture. Make the liquor quite warm, and put in as many shingles at a time as it will cover. When one batch of shingles is well soaked, remove and put in another. Then lay the shingles in the usual manner. With the liquor that is left, mix enough lime to make whitewash, and color with lampblack, ochre, or Spanish brown. Apply to the roof with a brush or old broom. This wash may be renewed from time to time.

SHINGLES, Painting.—Lay low-priced shingles—say from $2.75 to $4 per thousand—and paint them with a coat of tar and asphaltum—say one barrel coal-tar, costing $3; ten pounds of asphaltum at 3 cents, 30 cents; ten pounds ground slate, at 1 cent, 10 cents; two gallons dead oil at 25 cents, 50 cents, which should be added after the other has been wetted and thoroughly mixed. This mixture is as good as any thing that can be put on to shingles, as it will thoroughly keep the water out; and, if dry, they will not rot under the lap, nor will the nails rust.

SLATES, ROOFING, Selection of.—Dark purple and green slates are the best for roofing; others are liable to fade unequally, and produce a disagreeable appearance.

SLED-BODY, To build a transverse.—Make the sills out of 1-inch or ¾-inch boards, with cross-pieces of the same thickness bolted between the sills, which are double. You can make these very light and limber. Now put on your side-boards with a bolt down through the rave and sill, which will make it very stiff, and can be made very light, and with all the strength possible.

SMOKE-HOUSE, Cheap and good.—For 50 hams, make dimensions 7 x 8 feet. Dig all the ground out to below the frost line, and fill up to the surface with small stones. On this lay a brick floor with lime mortar. Walls of brick, 8 inches thick by 7 feet high, and a door on one side, 2 feet wide. Door of wood, lined within with sheet-iron. For the top, put on joists, 2 x 4, set up edgewise, and 8¼ inches from centre to centre, covered with brick and with a heavy coat of mortar. Build a small chimney over the centre, arching it over and covering it with a single roof in the usual way. An arch should be built on the outside, with a small iron door, similar to a stove-door, to shut it up. Make a hole in the arch through the wall of the house, and put an iron grate over it. The arch is much more convenient to put the fire in, than to build the fire inside the house, and the chimney causes a good draught through the latter. Burn good corn-cobs or hickory wood. This house should cost about $20.

STABLES, Building.—Bricks, built in hollow walls, are better than any other material. Commence with a stone foundation— the bottom course of which is broader than the stone-work above it—laid in half cement mortar up to the grade line, and then build the brick wall upon that, filling in all the space inclosed by the walls with concrete up to the line of the top of the water-table. Then pave with stones, firmly bedded to form a floor. On the outside, there should be a stone water-table, 8 or 10 inches high, projecting 1 or 2 inches outside of the main walls above, and having the upper surface of the projection beveled off to shed the water. Just above the water-table, it is well to have a course of slate built in the full thickness of the walls, which will prevent any dampness rising up into them from the ground by capillary attraction. Above the water-table, the walls should be built up with a smooth face, and with close, neatly struck joints inside as well as out, so as to present a clean, even surface, which should always be kept painted or washed with a lime or cement wash. Above the wall-plate, the space should be filled in to the under side of the roof-boards. The ceilings over the main story should be lathed and plastered ; partly for the sake of the neat appearance, partly to keep away cobwebs which infest exposed beams, and partly to prevent foul air rising from the room below, and tainting the hay-loft. The doors and windows inside should be trimmed with architraves, even if the latter be merely strips of the cheapest stuff. It may be desirable to fur out and lath and plaster the walls of a stable ; but if this is to be done, it is better to wainscot with wood up to the height of, say 5 feet, and to fill in the space between the walls and the wainscot, as high as practicable, with broken glass and mortar, and then to lath and plaster from the wainscot up to the ceiling. A wooden stable, too, may with advantage be treated in the same way, but the space behind the wainscot being wider, may be packed with bricks and mortar, and made solid in that way.

TIMBER, Strength of.—The strongest side is that which in its natural position faced the north.

TIMBER, To test the soundness of.—Apply the ear to the middle of one of the ends, while another person strikes upon the opposite extremity. If the wood is sound and of good quality, the blow is very distinctly heard, however long the beam may be. If the wood is disaggregated by decay or otherwise, the sound will be for the most part destroyed.

TROUGH, To make a tight.—Joint up the plank, and then, with a wide punch, set down a groove about $\frac{1}{16}$ in. deep the whole length ; then take off two or three shavings more, and put the trough together. When the wet gets into that joint the groove swells out again just the thickness it was at first, and of course two or three shavings thicker than the plank, and so closes all up tight. Wood can also be ornamented by punching down carefully in patterns, planing off a little, and then wetting ; the parts punched down show in relief above the planed surface, and make quite a puzzle.

WATER-CLOSET, To put up a.—The engraving represents sectional views of the water-closet in the upper floor of a two-story house. A A is the level of the surface of the ground at the back court and of the kitchen floor. B is a 6-inch vitrified fire-clay siphon-trap, with an open iron grating, C, at its top, which grating may be hinged. D is 4-inch soil-pipe from the water-closet ; it is here shown coming down inside the wall ; in other cases it may be carried down the outside. One advantage of such pipes being carried down the inside is that they are more likely to be protected from frost. F is a $\frac{1}{2}$-inch or 2-inch lead pipe for ventilating the soil-pipe. In this case, it is carried through the wall ; in other cases it may be carried up through the roof. G is the water-closet trunk, made of iron, it being a pan water-closet which is here shown. H is a $\frac{3}{4}$-inch lead pipe, carried through the wall, and put in to ventilate the trunk, or that space between the water in the pan, I, or basin, J, and the water in the siphon-trap, E. This $\frac{3}{4}$-inch ventilating pipe, H, is a very important one, and its use ought to be the rule in place of the exception, as is at present the case. It works as follows : When the handle of the water-closet is lifted, then any foul air lying in the trunk, in place of coming out into the apartment, is sent outside with a rush through this pipe, H ; besides, being open to the air, it tends to prevent the accumulation of such foul air in the trunk.

In order to keep the outer orifices of the pipes, F and H, always open, it is a good plan to solder on one or two pieces of copper wire across them. J is the water-closet basin, and the two small circles shown, underneath K K, are the india-rubber pipes. A 3-inch zinc ventilating pipe may be carried up through the roof to ventilate the space or inclosure in which the water-closet is situated. A gas-bracket placed right below it will help, when lighted, to cause an upward current. The empty space at N is

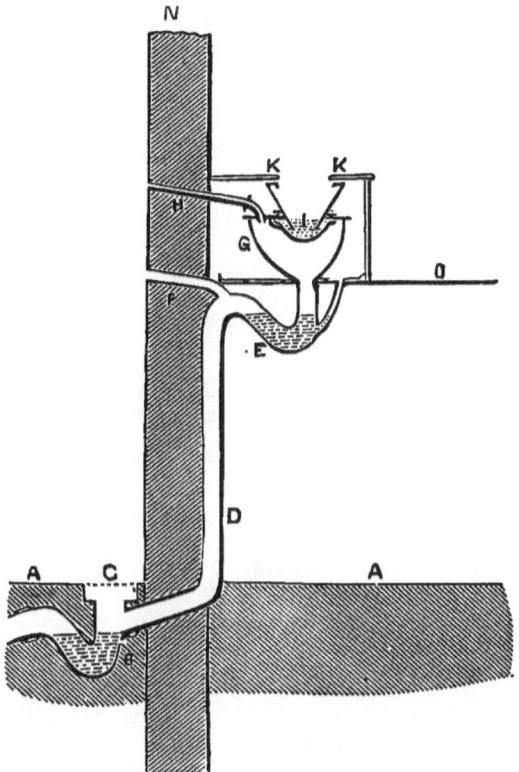

BUILDING WATER-CLOSETS.

supposed to be the water-closet window. O is the surface of the
floor of the upper flat. No gas can accumulate in the soil-pipe,
for the pressure of the atmosphere on the surface of the open
grating, C, tends to send a current of fresh air through the soil-
pipe and out at the ventilating pipe, F.

WATER-GATE, A good.—This is an excellent device for fencing
purposes over small streams. A gate, sliding in upright ways at
the ends, like an old turnpike gate, has attached to the bottom
board (a scantling is better, as not so likely to be broken in high
water) crutches which rest upon empty barrels or casks. The uprights at the ends of the gate are provided with friction-rollers,
so that the gate slides up and down easily in the ways. Two or
three casks will generally support the weight of the gate, so that
it descends nearly to the surface of, but does not enter, the water.
A gate thus constructed will rise and fall with the stream, and
is not liable to be washed away at high water.

WHITEWASH, for outside work.—Take quicklime, ½ bushel ;
slake, and add common salt, 1 pound ; sulphate of zinc. ½
pound ; sweet milk, 1 gallon. Dissolve salt and zinc before adding, and mix the whole to proper consistence with water.

WOODEN BUILDINGS, To frame.—Particular attention should
be paid to binding the top of the walls well together. This is
accomplished by framing the wall-plate all around the house,
and spiking the ceiling joists down on the same ; then herring-
bone, bridging these joists in as many rows as are necessary to
make a thoroughly stiff brace for the whole. The roof (no matter whether Gothic or Mansard) can not exercise any bad influence in pushing out the walls, when this system is adopted.

WOOD, To season and prevent warping.—Strip off bark, and
bury about one foot deep in the spring, leaving in the ground for
six months, and you will find no difficulty. This was the only
way by which the sapadillo or mountain mahogany in the Sierra
Nevada could be seasoned, it being one of the hardest and most
brittle kinds of wood known.

WINDMILL, To build a.—Windmills can be either horizontal
or vertical, but the latter are almost exclusively employed. In
the vertical windmill, the shaft is inclined to the horizon at an
angle of from 5° to 15°, when the wheel is placed at the top of a
tower ; so that the wheel will clear the sides of the building,
and allow space for the action of the wind. If the wheel is
supported by a post, the shaft may be horizontal. The connection of the shaft with the pump or other mechanism may be
made either with gearing or by means of a crank and connecting
rod. The shaft must be free to swing around in any direction,
so that the wheel can always face the wind. It is moved, in the
case of small windmills, by the use of a' weather-vane on the
end of the shaft opposite to the wheel. With large windmills
supported on towers, the top of the tower is generally arranged
so that it can be rotated, and a small auxiliary wind-wheel, connected by gearing, moves it into the proper position as the direction of the wind changes. The wheel of a windmill may be
covered with cloth, or with slats of wood or metal, the cover in
either case being technically known as the sail.

Make the sail of a series of joined slats, that present a close
surface to wind of the ordinary velocity, and open, thereby decreasing the surface, as the velocity of the wind increases. The
best velocity for a windmill is such that its periphery moves
about $2\frac{3}{5}$ times as fast as the wind. Thus, if the wind is moving
at the rate of 20 feet a second, the tips of the sails should move
at the rate of 52 feet a second, so that, if the wheel were 12 feet
in diameter, it should make about 83 revolutions a minute. Of
course, if the velocity of the wind varies greatly, it will be
impossible to keep the speed constant, so that windmills are not
ordinarily well suited for work requiring steady motion ; although
they answer very well for moving pumps, if an intermittent
supply of power is not a serious obstacle. In some sections,
however, the prevailing winds are quite steady, and in such
cases windmills can be applied with advantage to grist-mills
and other useful work. The force and velocity of the wind can
only be determined by experiment, but the results of previous
experimenters may be useful :

Velocity of wind.		Perpendicular force, in pounds per square foot.	Common expressions of the force of the wind.
In feet per second.	In miles per hour.		
10	6.82	0.33	Gentle pleasant wind.
20	13.64	0.91	Brisk gale.
30	20.56	2.04	Very brisk.
40	27.27	3.92	High wind.
50	34.09	6.25	Very high.
60	40.91	9.25	Very high.
70	47.73	12.75	A storm.
80	54.55	16.34	A storm.
90	61.36	20.74 ⎫	
100	68.18	25.28 ⎬	A great storm.
110	75.02	30.89 ⎭	
120	81.84	36.75	A hurricane.
130	88.65	43.26	A hurricane.
140	95.47	50.32	A violent hurricane.
150	102.29	57.56	A violent hurricane.

In the accompanying figure is shown one of the four sails of a windmill, it having been found that four sails of proper proportion produce the best effect. The piece P B is called the whip of the sail ; C D, E F, G H, etc., the bars of the sail. The bars are inclined to the plane of revolution at different angles, the angle made by any part of the sail with this plane being called the weather of the sail. Making the distances A O, N L, L I, etc., each equal to $\frac{1}{12}$ of the diameter of the wheel, the best values for the angle of weather are as follows:

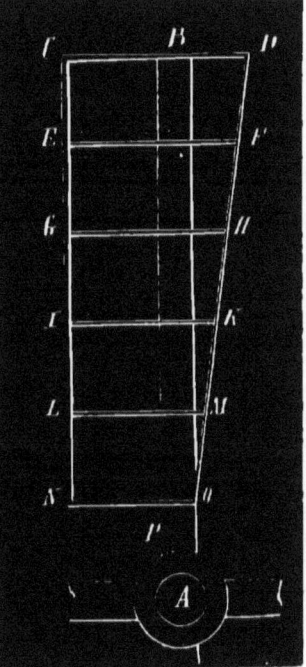

For N O—18°   For G H—16°
For L M—19°   For E F—12½°
For J K—18°   For C D—7°

The sail stretched over these bars will be a warped surface, somewhat resembling the blade of a screw-propeller. The part B D O, called the leading sail, is triangular, and B D is $\frac{1}{15}$ of the diameter of the wheel, B C being $\frac{1}{10}$, and C N $\frac{5}{12}$ of the diameter. The main body of the sail, B C N O, is commonly rectangular. A windmill of the best proportions, running under the most favorable circumstances, utilizes about $\frac{25}{100}$ of the energy of the

BUILDING A WINDMILL.

wind that acts on an area equal to a circle having the same diameter as the wheel. It would not be advisable to count on realizing more than half this power in general practice ; and on this assumption, we have the following empirical rule for determining the diameter of a wheel to give a certain amount of power with an assumed velocity of the wind :

Divide the required horse-power by the cube of the velocity of the wind in feet per second ; take the square root of the quotient and multiply it by the number 2024.8. The product will be the required diameter in feet. *Example :* A windmill is to be erected in a locality where the general velocity of the wind is about 20 feet per second. It is to be attached to a pump, the work required of it being to raise 1000 gallons of water per hour through a height of 20 feet   1000 United States gallons of water weigh about 8320 pounds, and, taking into effect the resistance of the pump, the power required will be about ⅙ of a horse-power, or 0.167 horse-power. Dividing this by 8000, the cube of the velocity of the wind, extracting the square root, and multiplying by 2024.8, we obtain 9¼ ft. as the required diameter of the wheel. Referring to the figure, we find that, in this case, C N is 3 feet 10¼ inches, B D, 7⅞ inches, and B C, 11¾₂ inches. The velocity of the tips of the sales should be 52 feet per second, or the wheel should make about 108 revolutions a minute.

THE DAIRY.

BUTTER, Philadelphia.—The pans containing milk to the depth of 3 inches are set in flowing water, so as to be maintained at a temperature of about 58° Fahr.   After standing 24 hours, the milk is skimmed, and the cream put in deep vessels of a capacity of about 12 gallons.   It is kept at a temperature of 58° to 59° until it acquires a slightly acid taste, when it goes to the churn.   The churn is a barrel revolving on a journal in each head, and is driven by horse-power.   The churning occupies about an hour ; and after the buttermilk is drawn off, cold water is added and a few turns given to the churn.   The water is then drawn off.   This is repeated until the water as it is drawn off is nearly free from milkiness.   The butter is worked with butter-workers, a dampened cloth meanwhile being pressed upon it to absorb the moisture and free it of buttermilk.   The cloth is frequently dipped in cold water, and wrung dry during the process of wiping the butter.   It is next salted at the rate of 1 ounce salt to 3 pounds butter, thoroughly and evenly incorporated by means of a butter-worker.   It is then removed to a table, where it is weighed out and put into pound prints.   After this, it goes into large tin trays, and is set in the water to harden, remaining until next morning, when it is wrapped in damp cloths and placed upon shelves, one above another, in tin-lined cedar tubs, with ice in the compartments at the ends ; and then it goes immediately to market.   A Philadelphia butter dealer says that, for the best butter, the cows are fed on white clover and early mown meadow hay, cut fine and mixed in with corn meal and wheaten shorts.   No roots are fed, except carrots.

CHURNING MILK, Temperature for.—60° Fahr.

CREAM-GAUGE.—In a can 20 in. deep and 8 in. in diameter, cut a slot a few inches long. In this slot, insert a strip of glass, in grooves, and cement with white lead. Graduate the tin next to the glass. Set the milk in the can, and allow the cream to rise. The percentage may be seen on the glass and noted by the scale.

CREAM, White specks in.—These are caused by too much acid in the cream. Cream should never stand in a room where the milk is set, but should be put into a cool place if you would avoid specks.

COWS, Care of.—Milk coming from ill-nourished, half-fed cows, having no surplus of food beyond minimum requirements of nature, is injurious, and may be a source of disease. Cows deprived of an abundance of good water, ventilation, and exercise, secrete impure and dangerous milk, which may be loaded with gases, animalculæ and fever-germs. The milk from old, debilitated cows fed on grains or overstimulating food, is also imperfect and unhealthy to a variable degree. The nervous condition of the cow at the time of milking determines the purity of the milk. If this is neglected, the milk is an active source of disease, and is positively dangerous and fatal.

COW-STABLES, Ventilation of.—Lay the floor of the stable upon a solid bed of earth and gravel, with a fall of 6 in. in 12 ft. from the stanchions, with the same ratio of descent, to a point for outlet of liquids. Make a platform raised 6 in. for the cows to stand or lie upon. The floor and platform plank should be bedded in water-lime mortar, so that there shall be no soaking down nor hiding-place for stale urine to deposit and generate venomous odors.

MILK, Poisonous sour.—Sour milk, after protracted exposure to the sun, develops a poisonous quality, sufficient to cause disease and death to pigs fed thereon.

MILK, Setting.—Place the pans in cold water, which will protect the milk from the acid until the cream has time to rise. For cream to rise readily on milk, set in cold water; the atmosphere in the room should be warmer than the water. There will as much cream rise on milk set in cold water in one hour as there will on milk not set in water in 24 hours.

MILK, Tainted.—Never allow dead animals to decay about a pasture, or any where near a barn or other localities inhabited by the milch-cows. The carrion odor is sure to affect the milk.

MILK, Testing for cream.—A simple method of determining the quantity of cream in any sample of milk consists in agitating the milk in a graduated glass tube with its bulk of ether for 4 or 5 minutes. Add alcohol in volume equal to that of the milk, and shake for 5 minutes. Place the tube vertically and allow it to rest for a brief period, when the oily matter will rise to the surface so that its amount may be read off on the scale and the percentage easily computed.

MILK, To insure good.—The following questions Mr. X. A.

Willard, a well-known dairy authority, recommends to be written out and posted about the dairy : " Do your cows feed in swamps and on boggy lands ? Have you good, sweet running water convenient for stock, and is it abundant and permanent in hot, dry weather ? Have you shade-trees in your pasture, or do you think that cows make better milk while lying down to rest in discomfort in the hot sun ? Do you use dogs and stones to hurry the cows from pasture at milking time, thus overheating their blood and bruising their udders ? Do you cleanse the udders of cows before milking by washing their teats with their own milk, and practice further economy by allowing their drippings to go into the milk-pail ? Do you enjoin your milkers to wash their hands thoroughly before sitting down to milk ? When a cow makes a misstep while being milked, do you allow your milkers to kick her with heavy boots, or to pound her over the back and sides with a heavy stool ? Is the air about your ' milk-barn ' or milk-house reeking with the foul emanations of the pig-sty, the manure-heap, or other pestiferous odors ?"

MILK, To prevent souring by thunder-storms.—A fire started in the dairy is an excellent preventive. This should be done even in the hottest weather. The object is to remove the damp, moist, heavy air, which is injurious to the milk.

MILK, To remove taste of turnips in.—Give the cow no turnips for two or three hours before milking. It is better to feed only the centre of the turnip, cutting off the top and bottom. A tablespoonful of nitre dissolved in as much water as it will take to a gallon of milk, placed in the pail before milking, is said to remove the taste of the vegetable.

FARM HINTS AND RECIPES.

BEE MOTHS, To kill.—Bee moths can easily be killed in large numbers by setting a pan of grease, in which is a floating ignited wick, near the hives after dark. The moths will fly into the light and fall into the grease.

BONES, Reducing.—Place them in a large kettle filled with ashes, and about one peck of lime to a barrel of bones. Cover with water and boil. In 24 hours all the bones, with the exception, perhaps, of the hard shin-bones, will become so much softened as to be easily pulverized by hand. They will not be in particles of bone, but in a pasty condition, and in an excellent form to mix with muck, loam, or ashes. By boiling the shin-bones 10 or 12 hours longer, they will also become soft.

BONES. Value of, as a fertilizer.—100 lbs. of dry bone-dust add to the soil as much organic animal matter as 300 lbs. of blood or flesh, and also at the same time $\frac{2}{3}$ their weight of inorganic matter—lime, magnesia, common salt, soda, and phosphoric acid. Superphosphate of lime. commonly used by farmers, is simply bones treated with ⅓ their weight sulphuric acid and an equal quantity of water.

CARBONIC ACID GAS, Removing from wells, cisterns, etc.—(1.) A bellows with a rubber hose reaching near the bottom will soon blow out the gas. (2) Let down a large bucket, draw up and empty the gas as if it were water. (3.) Pour down water; do this when a person falls to the bottom from inhaling the gas. (4.) Let down an umbrella spread, and pull up quickly several times in succession.

CATERPILLARS, Exterminating.—Orchard or tent caterpillars leave their rings of eggs on the young twigs. If these are cut off with a clipping pole, it will prevent in every instance a large nest of caterpillars, and be much more easily done than after the latter have grown.

CATERPILLARS, Remedy for.—A solution (1 part in 500) of sulphide of potassium, sprinkled on the tree by means of a hand-syringe, is extensively used in France.

CELERY, Propagation of.—A deep trench should first be dug, at the bottom of which a layer of sticks of wood, say 6 in. thick, should be placed, a drain-pipe being placed endwise upon one or both ends of the layer. The sticks should be then covered with about a foot of rich mould, wherein the plants should be set in a row, and about 5 in. apart. The plants should be well watered, the water being supplied through the drain-pipes, so that, passing through the layer of sticks, which serves as a conduit, the water is supplied to the roots of the plants. In earthing up, care should be exercised to close the stems of the plant well together with the hand, so that no mould can get between them. The earthing process should be performed sufficiently frequently to keep the mould nearly level with the leaves of the outside stems. If these directions are carefully observed, the plant may be grown at least 4 ft. in length, and this without impairing the flavor.

CHARCOAL, Effect of, on flowers.—All red flowers are greatly benefited by covering the earth in their pots with about an inch of pulverized charcoal. The colors (both red and violet) are rendered extremely brilliant. Yellow flowers are not affected in any way by charcoal.

CHICKENS, To fatten.—The best food for this purpose is Indian meal and milk.

CHINCH-BUGS, To destroy.—Put old pieces of rag or carpet in the crotches of the trees attacked. When the worms spin, as they will, in the rags, throw the latter in scalding water. The bugs can thus be killed by wholesale.

CIDER BARRELS, To clean.—Pour in lime-water, and then insert a trace-chain through the bung-hole, remembering to fasten a strong cord on the chain so as to pull it out again. Shake the barrel until all the mould inside is rubbed off. Rinse with water, and finally pour in a little whisky.

CIDER CASKS, To prepare.—Cider should never be put into new casks without previously scalding them with water containing salt, or with water in which apple-pomace has been boiled. Beer-casks should never be used for cider, nor cider-casks for beer. Wine and brandy casks will keep cider well, if the tartar adhering to their sides is first carefully scraped off and the casks be well scalded. Burning a little sulphur in a cask will effectually remove must.

CIDER from apple-parings.—The parings of a bushel of apples will yield 1 qt. of cider by the aid of the hand-press.

CIDER-MAKING, Hints for.—1. No good cider can be made from unripe fruit. The nearer to perfect ripeness the apples, the better the cider. 2. No rotten apples, nor bitter leaves, nor stems, nor filth of any kind should be ground for cider. 3. Two presses are really necessary for each mill, so that the pomace can be exposed to the air in the one, while it is being pressed in the other, and thus acquire a deeper color. 4. New oak barrels, or those in which whisky or alcohol has been kept, are the best. 5. If more color and richer body are desired, a quart or two of boiled cider to each barrel will impart them.

CIDER, Purifying.—Cider may be purified by isinglass, about 1 oz. of the latter to the gallon. Dissolve in warm water, stir gently into the cider, let it settle, and draw off the liquor.

CIDER, To preserve sweet for years.—Put it up in air-tight cans, after the manner of fruit. Rack it off the dregs, and can before fermentation sets in.

COAL-ASHES as a fertilizer.—Mix them with a small proportion of well-rotted horse-manure, sifting the ashes first, and you will have an excellent fertilizer.

CORN-COBS, Utilization of.—Save the corn-cobs for kindlings, especially if wood is not going to be plentiful next winter. To prepare them, melt together 60 parts resin and 40 parts tar. Dip in the cobs, and dry on sheet-metal heated to about the temperature of boiling water.

DOGS, Bed for.—The best is newly made deal shavings. They will clean the dog as well as water, and will drive away fleas.

DUST, ROAD, Value of.—During a dry season, every country resident should secure several barrels of road-dust. Those who keep poultry may secure by its use a valuable fertilizer, nearly as strong as guano, with none of its disagreeable odor. Place an inch or two of road-dust in the bottom of a barrel; then, as the poultry house is regularly cleaned, deposit a layer an inch thick of the cleanings, and so on, alternately layers of each till the barrel is full. The thinner each layer is, the more perfect will be the intermixture of the ingredients. If the soil of which the road-dust is made is clayey, the layers of each may be of equal thickness; if sandy, the dust should be at least twice as thick as the layer of droppings. Old barrels of any kind may be used for this purpose, but if previously soaked with crude petroleum or coated with gas-tar, they will last many years. If the contents are pounded on a floor into fine powder before applying, the fertilizer may be sown from a drill. Road-dust is one of the most perfect deodorizers of vaults—converting their contents also into rich manure. Place a barrel or box of it in the closet, with a small dipper, and throw down a pint into the vault each time it is occupied, and there will be no offensive odor whatever. This is simpler, cheaper, and better than a water-closet, and never freezes or gets out of order. Mixing the road-dust with equal bulk of coal-ashes is an improvement, making the fertilizer more friable.

FERTILIZER, A cheap.—This consists of sulphate of ammonia, 60 lbs. ; nitrate of soda, 40 lbs. ; ground bone, 250 lbs. ; plaster, 250 lbs. ; salt, ½ bushel ; wood ashes, 3 bushels ; stable manure, 20 bushels. Apply the above amount to six acres. Labor in preparing included, it costs about $15. It is said to give as good results as most of the commercial fertilizers costing $50 per ton.

FISH-NET, To preserve from decay.—Steep in melted paraffine.

FOWLS, Fattening.—It is said that charcoal will fatten fowls, and at the same time give the meat improved tenderness and flavor. Pulverize and mix with the food. A turkey requires about a gill a day.

FRUIT, To preserve.—Fruit is kept in Russia by being packed in creosotized lime. The lime is slaked in water in which a little creosote has been dissolved, and is allowed to fall to powder. The latter is spread over the bottom of a deal box, to about one inch in thickness. A sheet of paper is laid above, and then the fruit. Over the fruit is another sheet of paper, then more lime, and so on, until the box is full, when a little finely powdered charcoal is packed in the corners, and the lid tightly closed. Fruit thus inclosed will, it is said, remain good for a year.

GEESE AND GANDERS, To distinguish.—The goose has always a feminine appearance and the gander a masculine. Her head is smaller and her beak shorter ; knot on forehead smaller and not so pointed ; her neck shorter and more delicate ; the black streak on back of neck not so high ; colored ring around head not so bright ; her neck comes out of her body more abruptly (this is occasioned by her having a larger breast than the gander), giving a square appearance to the body. The voice of the gander is keener and louder ; coloring about head more brilliant ; eyes keener and always on the lookout.

GRAFTING WAX.—Take beeswax 1 part, resin 1 part, with enough tallow to soften. Melt all the ingredients together.

GRAFTS, Cutting and storing.—There is no better time to cut grafts, than at the commencement of winter. In cutting and packing them away, let them be labeled. For this purpose they should be tied up in bunches, not over 2 or 3 inches in diameter, with 3 bands around each bunch—at the ends and middle. The names may be written on a strip of pine-board or lath, ½ in. wide, ⅒ in. thick, and nearly as long as the scions. This, if tied up with the bunch, will keep the same secure. For convenience in quickly determining the name, there should be another strip of lath, sharp at one end, and with the name distinctly written on the other, thrust into the bundle with the name projecting from it. If these bunches or bundles are now placed on end in a box, with plenty of damp moss between them and over the top, they will keep in a cellar in good condition, and any sort may be selected, and withdrawn without disturbing the rest, by reading the projecting label. It is needful, however, to keep an occasional eye to them, to see that the proper degree of moisture is maintained—which should be just enough to keep them from shriveling, and no more.

GRAPES, To ripen.—In the Rhine district, grape-vines are kept low and as near the soil as possible, so that the heat of the sun may be reflected back upon them from the ground; and the ripening is thus carried on through the night by the heat radiated from the earth.

GRASSHOPPERS, To utilize.—The grasshoppers, desiccated and ground, are useful as a fertilizer; but in this prepared condition, they form an excellent food for all insect-feeding birds. There is no better food for all young domestic fowls. Containing silicic acid in a soluble state, they seem specially adapted for young birds, promoting the growth of feathers.

GRASSING A SLOPE.—A steep slope may be grassed over without sodding by first smoothing the surface and then mixing a tough paste or mortar of clay, loam, and horse-manure, with sufficient water. The grass seed, which should be a mixture of Kentucky blue grass and white clover, should be thickly but evenly scattered upon the moist surface of this plaster as it is spread upon the bank. The plaster should be at least one or two inches thick, and a thin layer should be laid over the seed. The surface should be kept moist, and a light dressing of some active fertilizer would help the growth. In a few weeks the growing grass should be cut, and should be kept short at all times until a thick sod is formed.

GUANO, Handling.—Many cases of poisoning have occurred by contact of guano with wounds. It should be handled with gloved hands.

GUANO, Home-made.—Make a compound of 1 bushel ashes, 2 bushes fowl-manure, 1¼ bushels plaster, and 4 bushels muck. Spread the muck on the barn-floor and dump the fowl-manure on top of it. Pulverize the latter with the spade, and mix in the other ingredients. Moisten the heap with water, or, better, with urine, before planting. Deposit about a handful in each hill of corn, potatoes, or beans, mixing it with the soil before putting in the seed.

HARNESS, Cleaning.—Unbuckle all the parts and wash clean with soft water, soap, and a brush. A little turpentine or benzine will take off any gummy substance which the soap fails to remove. Then warm the leather, and, as soon as dry on the surface, apply the oil with a paint-brush or a swab. Neat's-foot oil is the best. Hang up the harness in a warm place to dry, but do not let it burn.

HARNESS, Oiling.—Give one or two coats of lampblack and castor-oil warmed sufficiently to make it penetrate the stock readily. Then sponge the harness with 2 qts. warm soap-suds; when dry, rub over a mixture of oil and tallow, equal parts, with enough Prussian blue to give color. When well rubbed in, this compound leaves a smooth, clean surface.

HARNESS, Working team.—Do not use martingales on working teams. See that the hames are buckled tight enough at the top to bring the draft-iron near the centre of the collar. If too low, it not only interferes with the action of the shoulder, but gives the collar an uneven bearing.

HAY, Spontaneous combustion of.—Hay, when piled damp and in too large masses, ferments and turns dark. In decomposing, sufficient heat is developed to be insupportable when the hand is thrust into the mass, and vapors begin to be emitted. When the water is almost entirely evaporated, the decomposition continues, and the hay becomes carbonized little by little ; and then the charred portion, like peat—peat-cinders mixed with charcoal, sulphurous pyrites and lignite, etc.—becomes a kind of pyrophorus, by virtue of its great porosity and of the large quantity of matter exposed to high oxidation. Under the influence of air in large amount, this charcoal becomes concentrated on the surface to such a degree that the mass reaches a temperature which results in its bursting into flames. The preventives for this danger are care that the hay in the lofts is kept perfectly dry, that it is well stacked, and that it is stored in small heaps rather than in large masses.

HAY, To estimate the weight of.—Allow 1 cwt. of hay to the cubic yard.

HORSE, Power of.—The greatest amount an average horse can pull in a horizontal line will raise a weight of 900 lbs.; but he can only do this momentarily ; in continued exertion, probably half this amount is the limit.

HORSES, Bedding for.—Sawdust serves this purpose well.

HORSES, BUYING, Hints for.—Examine the eyes in the stable, then in the light ; if they are in any degree defective, reject. Examine the teeth to determine the age. Examine the poll or crown of the head, and the withers, or top of the shoulders, as the former is the seat of poll evil, and the latter that of fistula. Examine the front feet ; and if the frog has fallen, or settled down between the heels of the shoes, and the heels are contracted, reject him, as he, if not already lame, is liable to become so at any moment. Next observe the knees and ankles of the horse you desire to purchase, and, if cocked, you may be sure that it is the result of the displacement of the internal organs of the foot, a consequence of neglect of the form of the foot, and injudicious shoeing. Examine for interfering, from the ankle to the knees, and if it proves that he cuts the knee, or the leg between the knee and the ankle, or the latter badly, reject. "Speedy cuts" of the knee and leg are most serious in their effects. Many trotting horses, which would be of great value were it not for this single defect, are by it rendered valueless. Carefully examine the hoofs for cracks, as jockeys have acquired great skill in concealing cracks in the hoofs. If cracks are observable in any degree, reject. Also both look and feel for ringbones, which are callosities on the bones of the pastern near the foot ; if apparent, reject. Examine the hind feet for the same defects of the foot and ankle that we have named in connection with the front feet. Then proceed to the hock, which is the seat of curb, and both

bone and blood spavins. The former is a bony enlargement of
the posterior and lower portion of the hock-joint ; the second a
bony excrescence on the lower, inner, and rather anterior portion
of the hock ; and the last is a soft enlargement of the synovial
membrane on the inner and upper portion of the hock. They are
either of them sufficient reason for rejecting. See that the horse
stands with the front feet well under him, and observe both the
heels of the feet and shoes to see if he " forges " or overreaches ;
and in case he does, and the toes of the front feet are low, the
heels high, and the heels of the front shoes a good thickness,
and the toes of the hind feet are of no proper length, reject him ;
for if he still overreaches with his feet in the condition described,
he is incurable. If he props out both front feet, or points them
alternately, reject. In testing the driving qualities, take the reins
while on the ground, invite the owner to get in the vehicle first,
then drive yourself. Avoid the display or the use of the whip ;
and if he has not sufficient spirit to exhibit his best speed without
it, reject. Should he drive satisfactorily without, it will then
be proper to test his amiability and the extent of his training in
the use of the whip. Thoroughly test his walking qualities first,
as that gait is more important in the horse of all work than great
trotting speed. The value of a horse, safe for all purposes without blinds, is greatly enhanced thereby. Purchase of the breeder
of the horse if practicable ; the reasons are obvious.

HORSES, Common-sense treatment for.—(1) All horses must
not be fed in the same proportions, without due regard to their
ages, constitutions, and work. (2) Never use bad hay because it is
cheap ; there is no nourishment in it. (3) Damaged corn brings
on inflammation of the bowels and skin diseases. (4) Chaff is
better for old horses than hay, because they can chew and digest
it better. (5) Mix chaff with corn or beans. (6) There is not
sufficient nutritive body in either hay or grass alone to support a
horse under hard work ; in such case the food should chiefly be
oats. (7) For a saddle or a coach horse, half a peck of sound
oats and 18 lbs. of good hay is sufficient ; if the hay is not good,
add a quarter of a peck more oats. (8) Rack feeding is wasteful ;
feed with chopped hay from a manger. (9) Sprinkle the hay
with water that has salt dissolved in it, because it is pleasing to
the animal's taste and more easily digested. A teaspoonful of
salt in a bucket of water is sufficient. (10) Oats should be
bruised for an old horse, but not for a young one. (11) Cut grass
should always be given in spring to horses that can not be turned
out into the fields ; it is very cool and refreshing. (12) Water
horses from a pond or stream, rather than from a spring or well,
because the water from the latter is generally hard and cold, while
the former is soft and comparatively warm. The horse prefers
soft muddy water to hard water, though never so clear. (13) A
horse should have at least a pailful of water morning and evening, or (still better) four half pailfuls at four several times in
the day, because this assuages his thirst without bloating him.
He should not be made to work directly after a full draft of
water. (14) Do not allow a horse to have warm water to drink,
since, if he has to drink cold water after becoming accustomed to

warm, it will give him colic. (15) Do not work a horse when he refuses food after drinking ; he is thoroughly fagged out.

HORSES, Dead, To utilize.—Drag the body to some out-of-the-way part of the farm and sprinkle quicklime over it. Then cover with about twenty-five wagon-loads of muck or sods. In a year an excellent manure-heap will be at your disposal. Smaller animals may be similarly utilized.

HORSES, Scratches on.—These may be cured by washing the legs in warm, strong soap-suds, and then in beef brine.

HORSES, To keep flies from.—Make an infusion of 3 handfuls of walnut leaves in 3 qts. of cold water. Let this stand over night and boil for a quarter of an hour in the morning. When cold, rub it over the ears, neck, and other irritable parts of the animal, with a moist sponge.

HORSES, Wounds on.—If suppuration is inevitable, use carbolic acid combined with glycerine or linseed-oil in the proportion of 1 to 20. It may be applied night and morning with a feather. The wound must be kept clean, and, in the case of backs and shoulders, all pressure removed by small pads of curled horse-hair, sewed on the harness above and below the sore.

ICE, Compressed.—Thin ice from ponds, or small pieces left after cutting blocks from larger bodies of water, may be stored in a profitable manner, and at the same time its preservation insured, by compressing it into solid blocks by means of any simple press. In localities where ice is not attainable, snow might easily be treated in the same way.

ICE, To keep.—Select a shady spot, on the north side, if possible, of a clump of trees. Throw up a circular mound, some 12 in. in height and at least 15 ft. in diameter, flattening the summit carefully, and leaving a trench around the eminence, 2 ft. in width and 18 in. in depth. In gathering the ice, there is no necessity of cutting into uniform shape or of seeking large pieces. Fill up the carts with any kind of fragments, transport them to the mound, and dump them on a platform made of a few planks. Ram the surface of the mound hard and firm, cover with sawdust, and then place the first layer of ice, which should previously be cracked into small pieces, for which purpose the men should be provided with wooden mallets. As each layer is put on the stack, the ice should be thoroughly pounded, both above and at the sides, so as to form a huge block of ice, the shape of which will be slightly conical.

When the stack is completed, it will require two coverings of straw, one lying upon the ice and the other supported on a wooden framework about 18 in. outside the first covering.

The layer of straw next the ice must be well beaten and flattened down upon it, and when this is done, be 12 in. in thickness. The framework, upon which a similar thickness of straw is placed, may be formed by inserting stout larch or other poles of a suitable length round the base in a slanting direction, so that they can be readily brought together at the top, and securely fastened with stout cord. From six to eight of these will, when joined together by means of strips of wood fixed about 12 in. apart, afford ample support for the second covering of the straw. This must be put on nicely, so as to prevent the possibility of the rain's penetrating to the inner covering. By this arrangement there will be a body of air, which is one of the most effectual non-conductors known, between the two coverings of straw. To effect a change of the inclosed air, when rendered needful by its becoming charged with the moisture arising from the melted ice, a piece of iron or earthenware piping a few inches in diameter should be fixed near the apex, one end being just above the straw, and the other end reaching into the inclosed space. The pipe can be readily opened or stopped up, as may appear necessary, but as a rule it will suffice to open the pipe once a week, and allow it to remain open for about two hours. This should be done early in the morning, as the air is then much cooler than during the day or in the evening.

In removing ice from the stack, the early morning should be taken advantage of, because of the waste which must naturally ensue from a rush of warm air at midday. That removed can be placed in a cellar, or even an outhouse, and be enveloped in sawdust until required. The ice must be taken from the top ; and when the first supply is obtained, a good quantity of dry sawdust should be placed over the crown.

INSECT-CATCHING DEVICE, A simple.—Cover the inside of an old tub with liquid tar, and at twilight put a lighted lantern within, leaving the whole out over night. The bugs, attracted by the light, try to reach the lantern, and are caught and held fast by the tar.

INSECTS ON PLANTS, To discover.—If the leaves of the plant seem reddish or yellow, or if they curl up, a close inspection will generally disclose that the plants are infested by a very small green insect, or else with red spider, either of which must be destroyed. For this purpose, scald some common tobacco with water until the latter is colored yellow, and when cold, sprinkle the leaves of the plants with it. It is a good plan to pass the stems and leaves of the plants between the fingers, and to then shake the plant and well water the bed immediately afterward ; the latter operation destroys a large proportion of the insects shaken from the plant.

INSECTS, To protect cattle from.—Rub a very weak solution of carbolic acid through the hair.

MANURE, Salt as.—Salt should never be applied other than in a pulverous state, and never employed on impervious, cold, and humid soils. The best manner to use it is to combine it with other manures, a dose of 2 cwt. to the acre being sufficient. When selected to destroy insects, it should be applied before sunrise. In the case of cereals, salt strengthens the stems, and causes the ears to fill better, and favors the dissolution and assimilation of the phosphates and silicates. It acts vigorously on potatoes, and can be detected in their ashes to the extent of one half or one per cent. Asparagus is a veritable glutton in the presence of salt. A dose of 3 cwt. per acre acts without fail on beet, injuring its value for sugar purposes, but enhancing it for the feeding of cattle. Colza has as marked a predilection for salt as asparagus ; and in Holland, where the culture of peas is so extensive, salt is something like a necessity. Mixed with hay in the proportion of 4 ozs. to 1 cwt., the fodder is more appetizing ; but the best way to feed it to animals is to allow them to enjoy it in the form of rock salt.

MANURE, Soot.—Save the soot that falls from the chimneys, when the latter are cleaned. Twelve quarts of soot to a hogshead of water makes a good liquid manure, to be applied to the roots of plants.

MANURE.—The bodies of king crabs, often found along the seashore, when decayed and mixed with sawdust, straw, muck, or similar material, make an excellent manure. Land so poor that no wheat could be grown on it, has been so enriched by the application of this compost, that from 25 to 30 bushels to the acre have been raised.

MOSS ON TREES, To destroy.—Use a whitewash of quicklime and wood-ashes.

MOWING MACHINES, Draft of.—The power required to drive a mowing machine at work may be resolved into direct draft and side draft. A good mowing machine should be so balanced between the driving wheels and the cutter bar, by placing the line of draft nearer or further from the heel of the bar, that ordinarily there will be little or no side draft. If so placed, the end of the tongue will sometimes, when at work, be drawn toward the standing grass, and sometimes thrown away from it by the side draft. Practically, side draft is of small account in a good machine. The direct draft depends upon three causes, and may be resolved into three parts : 1, The draft of the machine itself, on its own wheels, on the ground ; 2, The power required to give motion to the gearing and the knife ; 3, The resistance offered by the grass or other substance cut. The power required to draw the machine on its own wheels depends upon the size of the wheels, the perfection of the axles, and the smoothness of the ground, and, other things being equal, upon the weight of the machine ; and in a machine weighing 600 lbs., should not, on a smooth firm turf and level field, be more than 75 to 100 lbs. Adding to the weight of the machine would add to the draft just in proportion, whether that added weight be in the machine or in a driver on it. Suppose the machine to weigh 600 lbs. and the draft to be 80 lbs., put a driver of 150 lbs. on the seat, and you have increased the draft 25 per cent, or to 100 lbs., while the power required to drive the knife and to cut the grass remains unchanged, and you have only increased the draft 20 lbs. The power to drive the knife and gearing depends upon the perfection of the gearing and the weight and velocity of the knife. A machine that in proper order may not require more than 10 or 20 lbs. of power, may require ten or twenty times that amount from deficiency of oil, collection of dirt in the gearing, gum on the knife, or loosening of the connections of the pitman by wearing or otherwise. The resistance of the grass to be cut will vary with every change of condition, kind, and thickness of grass, and every variation in the condition of the knife and rapidity of stroke. The greater the velocity of a cutting edge, after the velocity is once obtained, the less power is required to do a given amount of work. The direct draft of a good machine, working under the most favorable circumstances, has been determined by experiment to be less than 300 lbs. ; but those favorable conditions are not always to be obtained, so that the draft must many times be much heavier. The power of a machine to cut, other things being equal, depends upon the hold the wheels have upon the ground ; when the second and third causes combined are sufficient to overcome the hold the wheels have, the latter slide, the knife stops, the machine is clogged. The heavier the machine, the less likely this is to occur ; putting a heavy driver on the seat will sometimes carry a machine through when with a lighter one it would clog. When the knife reaches the end of the stroke, its momentum is considerable, and it requires nearly as much power to stop it as it did to start it to make the stroke ; it would require quite as much if it were not for the loss of some power by the friction of the knife in the guards. Now if the joints of the pitman and connections are all perfect, this stopping occurs when the crank passes the centre of the shaft driving it, and the remainder of the momentum of the knife is expended upon the crank in the direction of its length and at right angles to the driving power, so that none of that is used up in stopping the knife. But if these joints of the pitman are loose, so that there is a little play, and the crank can pass the centre before the knife reaches the end of the stroke, this momentum will be expended in opposition to the driving power, and will of course increase the power necessary to work the machine by so much as is necessary to overcome the momentum of the knife ; again, the crank beginning to act upon the knife after it has passed the centre to make the return stroke, the knife must start with a greater velocity, causing another loss. Now, put the same machine into the grass, and the grass operates to stop the knife as soon as the crank allows it to stop, thus saving the momentum that was expended upon the crank in opposition to the driving power, and also shortening the stroke and saving power that way. Supposing, to illustrate, that there is a play of ¼ inch in the joints, then the knife running out of the grass will be thrown to the extreme length each way, and will add 1 inch to the length of the stroke, increasing the power necessary to make it. If it requires a certain number of pounds power to make a stroke of 3 inches in length, it will require 33⅓ per cent more power to make a stroke of 4 inches in length in the same time. Whenever these amounts of power lost in this way equal the power required to cut the grass, then the machine will draw just as heavily out of the grass as in it. From these premises many deductions may be made as to the care and practical use of mowing machines.

NEST EGGS, To make.—Take an ordinary hen's egg, break a small hole in the small end, about ⅜ of an inch in diameter, extract the contents, and, after it is thoroughly clear inside, fill it with powdered slaked lime, tamping it in order to make it contain as much as possible. After it is full, seal it up with plaster of Paris, and you have a nest egg which can not be distinguished by the hen from the other eggs, and one which will not crack (like other eggs) by being frozen.

ONIONS, To transplant.—Plant them tightly in the soil, with the leaves pointing to the north. When thus placed, after the vegetables take root, the sun will draw the stalks vertical.

OXEN, Food consumed by.—An ox will consume 2 per cent of his weight of hay per day to maintain his condition. If put to moderate labor, an increase of this quantity to 3 per cent will enable him to perform his work and still maintain his flesh. If he is to be fatted, he requires about 4½ per cent of his weight daily in nutritious food.

PAINTING BOATS.—Boats should be painted with raw oil. Boiled oil used in the paint is very apt to blister and peel from the wood.

PARIS GREEN, To use.—In using Paris green to exterminate the potato bugs, the poison should be mixed with the cheapest grade of flour, 1 lb. green to 10 lbs. flour. A good way of applying it to the plants is to take an old 2 quart tin fruit-can, melt off the top, and put in a wooden head in which insert a broom-handle. Bore a hole in the head, also, to pour the powder in, and then punch the bottom full of holes about the size of No. 6 shot. Walk alongside the rows, when the vines are wet with dew or rain, and make one shoot at each hill.

PASTURES, Seeding.—Select varieties of seeds that spring up in succession, so that a good fresh bite may be had from spring to fall.

PEAR CULTURE.—Pears have a tendency to crack when the trees stand in soil which is deficient in lime and potash. Common wood-ashes contain these salts nearly in the proportions that pear-trees on such soil require—40 per cent of potash and 30 per cent of lime. By applying wood-ashes at the rate of four hundred bushels to the acre, after the fruit had formed and cracked, the disease was totally eradicated by the next season.

PEAT, Estimating quantity of.—Peat, as ordinarily in the bed, will weigh from 2100 to 2400 lbs. per cubic yard; and if drained in the bed, 1340 to 1490 lbs.; and air-dried, 320 to 380 lbs., when it will be found to be reduced to about ¼ or ⅙ its original bulk.

PEAT, Facts concerning.—When saturated with salt water, peat is generally unfit for heating purposes. The fine, clay-like powder found underlying peat-beds, of a yellowish white color, is composed of shields of infusorial animalculæ, and forms a superior powder for polishing metals. In working a bed of peat, the first step will be to ascertain if drainage is necessary; and, secondly, how it can be effected and at the least cost. If the bed can not be economically drained, resort must be had to mechanical excavation. It is best not to drain a bed below the level to which you can effectually work out in a season, unless you can close the outlet drain to allow it to fill again with water for the winter, for the reason that drained peat that has been frozen is apt to disintegrate after thawing, and become impoverished for a solid homogeneous fuel. Peat that has been well manipulated and dried for fuel rarely holds more than 10 per cent of moisture, and it will not afterwards become saturated with water, even by immersion for an entire winter. A cubic yard of closely-packed

peat fuel will weigh from 1620 to 2180 lbs., and the heating value
of 1 lb. of such peat is equal to even 1¼ lbs. wood ; one cord of
good wood will weigh almost 4200 lbs., and 1 cord of peat fuel
will weigh about 3750 lbs., showing a gain in space as well as in
greater heating power.

PLANTS, CREEPING, Pegging down.—To propagate lobelias and
verbenas, the first bloom should be picked off, and the branches,
as they extend, should be pegged down closely to the surface of
the mould. The branches will then take root as they lengthen,
and by thus drawing a large amount of sustenance from the soil,
they will bloom very freely and cover a large space. A verbena
may thus be made to cover a square yard, and a lobelia a square
foot of ground.

PLANTS, Iron and ashes for.—White flowers, or roses, that have
petals nearly white, will be greatly improved in brilliancy by
providing iron sand and unleached ashes for the roots of growing
plants. Ferruginous material may be applied to the soil where
flowers are growing, or where they are to grow, by procuring a
supply of oxide of iron, in the form of the dark-colored scales
that fall from the heated bars of iron when the metal is hammered by the blacksmiths. Iron turnings and iron filings, which
may be obtained for a trifle at most machine-shops, should be
worked into the soil near flowers, and in a few years it will be
perceived that all the minute fragments will have been dissolved, thus furnishing the choicest material for painting the gayest
colors of the flower-garden. If wood-ashes can be obtained readily, let a dressing be spread over the surface of the ground, about
half an inch deep, and be raked in.

A dressing of quicklime will be found excellent for flowers of
every description. It is also of eminent importance to improve
the fertility of the soil where flowers are growing, in order to
have mature, plump, ripe seed. Let the foregoing materials be
spread around the flowers, and raked in at any convenient period
of the year.

PLANTS, Potting.—The mould for potting should be light and
loamy, the fertilizing material used being well decayed. If the
soil is rich of itself, it is better to be either very sparing with the
fertilizer or to dispense with it altogether. In the bottom of the
pot place several small broken pieces of crockery or similar material to assist the drainage ; and in setting the plant, be careful
to keep it well down in the pot, and to press the mould moderately
around the roots. The surface of the mould should be about half
an inch below the level of the top of the flower-pot. Slips should
be planted close to the sides of the pot, and in small pots. When
a plant becomes pot-bound, that is, when the roots have become
matted around the sides and bottom of the pot, the plant, so soon
as it has ceased blooming, should be re-potted in a larger pot. It
is not necessary to remove any of the mould from the roots, but
simply to fill in the space in the larger pot with new and rich
mould.

PLANT-PROTECTOR, A newspaper.—A convenient number of
newspapers may be pasted together, and the edges folded over
strings, thus making a screen which, suspended over the newspa-

pers spread loosely over the plants, would give the young shoots an excellent protection in the severest cold weather, and from the sun's rays in summer.

PLANTS, Selecting.—Choose those whose leaves are of a deep green, and in all cases those which are short and bushy, and have no bloom upon them. If, however, they are in bloom, cut off the flowers before planting, which will only delay the blooming a few days, and will greatly strengthen the plant. If the plants have been reared in a greenhouse or under frames, keep them a few days before setting them in beds, placing them out of doors in the daytime, and taking them in at night, in order to make them hardy and prevent them suffering from the cool night air. If the plants are placed in a cold-frame, either before or after being planted in the beds, be careful to lift the frame during a great part of the daytime, otherwise the moisture which gathers on the inside of the glass will fall upon the plants and infallibly kill them by what is called dampness.

PLANTS, Treating unhealthy.—Mr. Peter Henderson, the celebrated horticulturist, says : Whenever plants begin to drop their leaves, it is certain that their health has been injured either by over-potting, over-watering, over-heating, by too much cold, or by applying such stimulants as guano, or by some other means having destroyed the fine rootlets by which the plant feeds, and induced disease that may lead to death. If the roots of the plant have been injured from any of the above-named causes, let the soil in which it is potted become nearly dry ; then remove the plant from the pot, take the ball of soil in which the roots have been enveloped, and crush it between the hands just enough to allow all the sour outer crust of the ball of earth to be shaken off ; then re-pot in rather dry soil (composed of any fresh soil mixed with equal bulk of leaf-mould or street-sweepings), using a new flower-pot, or having thoroughly washed the old one, so that the moisture can freely evaporate through the pores. Be careful not to over-feed the sick plant. Let the pot be only large enough to admit of not more than an inch of soil between the pot and ball of roots. After re-potting, give it water enough to settle the soil, and do not apply any more until the plant has begun to grow, unless, indeed, the atmosphere is so dry that the moisture has entirely evaporated from the soil ; then, of course, water must be given, or the patient may die from the opposite cause—starvation. The danger to be avoided is in all probability that which brought on the sickness, namely, saturation of the soil by too much water.

POTATOES, Hoeing.—By drawing up the earth over the potato in sloping ridges, the plant is deprived of its due supply of moisture by rains, for when they fall the water is cast into the ditches. Further, in regard to the idea that, by thus earthing up, the number of tubers is increased, the effect is quite the reverse ; for experience proves that a potato, placed an inch only under the surface of the earth, will produce more tubers than one planted at the depth of a foot.

POTATO SPROUTS, Poisonous.—The sprouts of the potato contain an alkaloid, termed by chemists solanine, which is very poisonous if taken into the system. It does not exist in the tubers unless they are exposed to light and air, which sometimes occurs from the accidental removal of the earth in cultivation.

POTATOES, Storing.—A plan, tested successfully for eight years, is to sprinkle the floor with fine unslaked lime, over which a layer of potatoes 4 or 5 inches in depth is spread. Then sprinkle again with lime, and add another layer of potatoes the same depth as before, and thus continue till the whole are disposed of. The lime used is about one fortieth part by measure of the potatoes.

Potatoes thus treated have never become infected with disease, and when disease was already existing it has not spread ; besides which, the quality of the potatoes has been rather improved than otherwise by the treatment, especially where they were watery or waxy.

POULTRY-HOUSES, Purifying.—Lime is an excellent purifier, and, when carbolic acid is added to the whitewash, will effectually keep away vermin from the walls. After every cleaning of the floor it should be sprinkled with carbolic acid ; dilution, twenty of water to one of acid. This is one of the best disinfectants and antiseptics known, and is not used as much as it deserves. The roosts should be sprinkled with it every week. This whitewashing should be done twice at least, better three times a year. The nests of sitting hens should be sprinkled with carbolic acid to keep off vermin ; and the coops also, where young broods are kept for a time, should be purified in this way. If a hen gets lousy, the dilute acid will destroy the lice, if put under the wings, and on the head and neck. Wood-ashes are excellent to be kept in fowl-houses for hens to dust themselves with. They are much more effectual than sand, but sand should also be kept for a bath.

PROPAGATING PLANTS.—To propagate geraniums and calceolarias, do not let the plants flower too soon, but pinch off the first appearing bloom, and pinch out the eyes of all straggling branches, which will immediately throw out side-shoots, thus forming very healthy and strong as well as good-shaped plants. Give preference to those plants which have their branches close to the surface of the soil.

PRUNING TREES.—The proper cut is called the " clean cut," and is made by cutting at an angle of 45°, beginning at the back of the bud, and finishing slightly above it. When pruned in this way the wound readily and rapidly heals, and commences to be covered with new wood as soon as the young bud pushes into growth. Pruning should always be done with a keen-edged knife, holding the shoot in the left hand, and making one sharp, quick draw. The operation should be delayed until the middle of February, and performed between that time and the middle of March.

RATS, Extermination of, by bisulphide of carbon.—Insert a lead pipe into the holes, and pour in bisulphide of carbon. This* should only be used out of doors, never in buildings. An ounce and a half of the liquid is sufficient to pour in at a time. Where there are several holes near together, stop all but the one in which the bisulphide is poured, with bricks.

RED SPIDERS, To exterminate.—Syringe the plants freely with water once or twice a day, taking care to wet the under side of the leaves. Keep the air of the room moist, by setting pans of water on the flues, heating-pipes, or register; give all the light possible, and ventilate freely whenever the weather will permit. When the soil is dry, give sufficient water to moisten all the soil in the pot, and water no more until the surface is dry again. If plants seem stunted or sickly, repot them in fresh, rich soil, or use some other means to induce a healthy growth. The red spider is any thing but an aquatic insect, and will yield to the hydropathic treatment if it is persisted in.

SAND is the best substance in which to preserve carrots through the winter. It should be perfectly dry. It will keep the roots crisp and prevent softening.

SPAWN-CARRYING DEVICE.—The apparatus represented here-

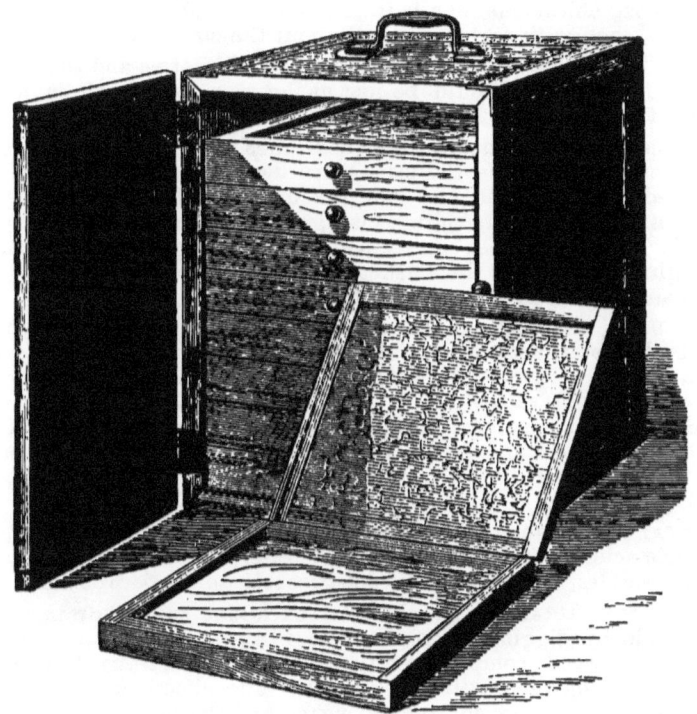

FISH-SPAWN CARRYING APPARATUS.

with is a new invention of Mr. Seth Green. It consists of a simple wooden box, of a convenient size to be carried in the hand by means of the handle above. Its joints are covered with tin. Inside are numerous small trays made of wood, covered below with canton flannel. The upper tray, shown in the foreground, is provided with a hinged cover of the same materials. The spawn is

placed upon the bottom of the trays, together with moss or sea-
weed, and kept moist. The temperature of the room may be so
regulated that the spawn can be hatched in from 50 to 150 days.
Brook-trout, salmon trout, white-fish, and salmon-eggs have been
transported with success, over long journeys, by this means.

SEEDS, Germination of.—The germination of seeds can be
watched at every stage of its progress by laying the seeds between
moist towels and placing the latter between plates. The towels
can be lifted without damage to the tender sprouts.

SEEDS, Vitality of.—Seed will not germinate if they are too
old, and disappointment and delay often result. Experience of
seedsmen indicates that, if properly gathered and preserved,
beans will retain vitality 2 years ; beet, 7 ; cabbage, 4 ; carrot, 2 ;
sweet corn, 2 ; cucumber, 10 ; lettuce, 3 ; melon, 10 ; onion, 1 ;
parsnip, 1 ; peas, 2 ; radish, 3 ; squash, 10 ; tomato, 7 ; turnip, 4.

SHEEP, To protect from dogs.—An old sheep-raiser says that
the most efficacious plan is to provide 15 or 20 sheep, in a flock
of 100, each with a globular bell about the size of a teacup.

SLEIGH.—The length of the double whiffletree and the neck-
yoke for a sleigh should be just as long as the sleigh is wide
from the centre of one runner to the other.

SLUGS, ROSE, To destroy.—Wood-ashes must be sifted on early
in the morning while the leaves are damp, the branches being
turned over carefully, so that the under sides of the leaves, to
which the young slugs cling, may get their share of the siftings.
If the night has been dewless, in order to make the work thorough, first sprinkle the bushes, and the ashes will then cling to
the slugs, to their utter destruction.

STUMPS, Clearing off.—In the autumn, bore a hole 1 to 2 inches in diameter, according to the girth of the stump, vertically in
the centre of the latter, and about 18 inches deep. Put into it
from 1 to 2 ozs. saltpetre ; fill the hole with water, and plug up
close. In the ensuing spring, take out the plug, pour in about ¼
gill kerosene oil and ignite it. The stump will smoulder away,
without blazing, to the very extremity of the roots, leaving nothing but ashes.

STABLES, To remove rank smell of.—Sawdust, wetted with sulphuric acid, diluted with 40 parts of water, and distributed about
horse-stables, will remove the disagreeable ammoniacal smell.

SUB-SOIL DRAIN, A simple.—An excellent subsoil drain may
be made by digging a trench, and filling in the bottom with
sticks of wood, compressing them together with the feet, and
then covering them with the mould. The effectiveness of such
a drain will endure for several years, and the final decay of the
wood will serve to enrich the soil.

SUMAC, Cultivation and preparation of.—Sumac is largely used
in tanning the finer kinds of leather, especially in the manufacture of the hard-grained moroccos and similar goods. It is also
employed as the base of many colors in calico and *delaine* print-
ing. The only trouble is in curing it properly. This must be
done with all the care that is bestowed upon tobacco or hops.
Exposure, after cutting, to a heavy dew injures it, and a rain-

storm detracts materially from its value. It is cut when in full leaf ; and when properly dried is ground, leaves and sticks together. An acre in full bearing will produce not less than three tons ; and when fit for market, it is worth from eighty to one hundred dollars a ton. The manufacturers, as the curers are called, pay one cent a pound for it in a green state. The Commissioner of Agriculture advises to plant in rows, in order to cultivate between, either by seed or cutting of the roots. We should advise cuttings by all means, as sumac is as tenacious of life as the blackberry or horse-radish. It will never need but one planting, and the crop can be gathered any time from July to the time of frost. If it is cut later in the season, and annually, the leaves and the stocks can be ground together. If the cutting is delayed until the stock has formed into solid wood, the leaves must be stripped from the stock, and the stock is thus wasted. It is doubtful if any thing is gained in the weight of leaves after the middle of July, at which time almost every tree has completed what is called first growth for the season.

SUMAC, Mordants for dyeing with.—The mordants used for dyeing with sumac are either tin, acetate of iron, or sulphate of zinc. The first gives yellow, the second gray or black, according to strength, and the third greenish-yellow.

TOOLS, Paint for.—White lead ground in oil, mingled with Prussian blue, similarly prepared, to give the proper shade, and finally mixed with a little carriage-varnish, is an excellent and durable paint for farm-machinery and agricultural tools.

TREES, Felling.—To find the height at which a tree must be cut, so that its top will strike a given point on the ground : Square the height of tree, and the given distance from tree to point. Divide the difference of these squares by twice the height of tree, and the quotient will be the height from the ground where the tree has to be cut. Example : Height of tree=60 feet, distance of point to the tree 20 feet ; then $60^2=3600$, $20^2=400$, difference=3200. $3200\div(2\times60)=26.6$ feet.

TREES, FRUIT, To protect against mice.—Apply to the bark a mixture of tallow, 3 parts, tar, 1 part, hot.

TREES, GIRDLED, To save.—If possible, bank up earth about the trunk to above the level of the girdle. If the wounded parts are too high, bind on clay with a bandage. The sooner the surface is protected after injury the better. The death of the tree is caused by the seasoning of the sap-wood.

TREES, YOUNG, in hot weather.—If the trunk is fully exposed to the sun, it should be protected from intense heat. A couple of boards, tacked together like a trough and set up against the trunk, will furnish the required shade ; or the trunk may be bound with a hay-rope, or be loosely strawed up as for winter protection.

TRICHINÆ IN PORK, A cause of.—It has lately been found that swine may become infested with trichinæ through eating carrion, or even decayed vegetable substances. This is a point worth consideration by farmers who incline to the belief that dead

chickens, putrid swill, or any other filth about the place, is legi-
mate food for the pig.

TURNIPS, To protect from fly.—Use lime, slaked just before
application. Sow it by hand over the plants ; or sow brimstone
with the seed. A simple way of removing the insects from the
plants is to mount a board two feet square on wheels, cover the
under side of the board with tar, and straddle it over the rows,
drawing it from end to end of the latter. The insects will jump
off on the pitch and be caught.

WATERFALL, To determine the height of, in a running stream.
—A small temporary dam, unless one exists, must be made, so
as to secure a still surface. Take two poles, sufficiently long to
reach from the bottom of the water to the required line-level.
Make a plain mark or notch on both sticks, at a distance from
the upper end equal to the distance of the intended line-level
above the water, marking that distance in feet and inches. Push
the poles down through the water into the earth at the bottom until
the notches are both at the level surface of the water, care being
taken to have the poles plumb and at a convenient distance apart.
Sight across the tops of these two, and set as many more as may
be desired to run the line of level to the desired point, and the
tops being ranged accurately by the first two, will show a water-
level so many feet above that of the water. It is estimated that
this is a more accurate way than the use of the ordinary spirit-
level.

WATER FOR FISH-PONDS.—Lead-pipe will not do to conduct
water to fish-ponds. It is likely to poison the fish.

WEEDS, Destroying.—Some weeds can be killed and prevented
from growing in garden-paths, by watering the ground with a
weak solution of carbolic acid, 1 part pure crystallized acid to
2000 parts water. Sprinkle from a watering-pot.

WORMS, CURRANT AND GOOSEBERRY, Remedy for.—The best
is powdered white hellebore, obtainable at any druggist's. Put
the powder in a common tin cup, tying a piece of very fine muslin
over the mouth. Fasten the apparatus to the end of a short
stick, and dust the powder through the muslin lightly upon the
bushes. Do not work on a windy day, and stand to windward
during the operation, as, if taken into the nostrils, the hellebore
excites violent sneezing. The same material is a good remedy
for cucumber-beetles.

HOUSEHOLD HINTS.

ANTS, RED, To drive away.—Sprigs of wintergreen or ground ivy will drive away red ants ; branches of wormwood will serve the same purpose for black ants. The insects may be kept out of sugar-barrels by drawing a wide chalk mark around the top, near the edge.

AQUARIUM, To make and stock an.—One of the first principles, in constructing a tank for an aquarium, is to give the water the greatest possible exposure to the air. The simple rectangular form is the best. This is generally constructed of iron and glass ; the iron should be japanned, and the glass be French plate, to insure brilliancy and strength. The breadth and height of the tank should be about one half of the length. Cheap tanks can be made of wood and glass, the frame and bottom being of wood, - and the sides of glass. In order to make the joints watertight, care must be taken to get a proper aquarium putty or cement. The following is a good recipe : Put an eggcupful oil and 4 ozs. tar to 1 lb. resin ; melt over a gentle fire. Test it to see if it has the proper consistency when cooled ; if it has not, heat longer or add more resin and tar. Pour the cement into the angles in a heated state, but not boiling hot, as it would crack the glass. The cement will be firm in a few minutes. Then tip the aquarium in a different position, and treat a second angle likewise, and so on. The cement does not poison the water. It is not advisable to make the aquarium of great depth ; about eight inches of water is sufficient. In regard to the light, great care must be taken. Too much often causes blindness, and is a common source of disease. The light fish receive in rivers comes from above, and an aquarium should be constructed so as to form no exception to this rule. All cross-lights should be carefully avoided, at least if the light is very strong. Never place the aquarium in front of a window so that the light passes through it ; for, when viewing an aquarium, the source of light should come from behind us. Not enough light is as injurious as too much, and causes decay of the vegetation. Having constructed a watertight aquarium, the bottom is strewn over with clean sand to the depth of 1 to 3 inches ; on this a little gravel is spread ; then a few stones or rock-work. Heavy large rocks should be avoided ; they displace a large amount of water, and increase the danger of breaking the glass sides. Pumice-stone, well washed, is the best kind, being light and with a rough surface suitable for the rooting of plants, etc.; and if fancy forms are desired (bridge-work, etc.), the pumice-stone can be cut quite easily to the desired shapes. The plants are rooted in the sand and the vessel left at rest for a week for the plants to vegetate. The following plants will be found useful : *Utricularia inflata, utricu-*

laria vulgaris, myriophyllum spicatum, anarcharis Canadensis, and *hottonia inflata.*

In obtaining plants, procure all the roots and see that they are well rooted. If fungus should form, add snails (*planorbis trivolvis*) ; they will completely destroy it. After the plants are well started, add the shells and amphibious animals. The following shells will be found desirable : *Planorbis trivolvis, physa heterostrapha, unio complanatus.* Many shells are not needed. Snails act the part of scavengers ; and where the different elements of an aquarium are rightly balanced, two or more snails will be found sufficient.

If amphibious animals are introduced, the rock-work must extend above the surface of the water, or a float of some kind must be substituted. It is impossible for them to live under water all the time, and they would die without some such arrangement.

The turtles claim first rank. The *enys punctata,* or spotted water-turtle, and the *chrysemys picta,* or painted water-turtle, will be found to be the best for the aquarium, and should be procured when very young, as they are very destructive when old. The tritons (*triton tigrinus, triton niger*), the red salamander, the cray-fish (*astacus Bartoni*), are all suitable, and present a very odd and yet a very natural look to the aquarium.

In selecting the fishes, a few only thrive in confinement. Among these, and the first, is the gold-fish. He can live for months without introduced food, and is, without comparison, the most hardy, standing remarkable changes in the temperature ; and he is the most gaudy and attractive. A large number of the fishes prey upon each other, and will only do for the aquarium when in the young state. Among these may be mentioned *pomotis vulgaris,* or sun-fish, *esox reticulatus,* or common pickerel, and *perca florescens,* or yellow perch. The *leuciscus pygmæus,* or rock-fish, is a great addition, and is found very plentifully in our streams. The *pimelodus atrarius,* or common black catfish, is another worthy of a place. So also is the *hydrargia diaphana,* or transparent minnow. But few fish can live in an aquarium, and the needless crowding together, so often seen, is very hurtful to health, and causes sound, strong fish in a short time to become weak and poor. The great difficulty in keeping an aquarium is to secure enough oxygen for the fish. To a slight degree, it is the duty of the plants to supply this ; but if too much vegetation be present, decomposition takes place and ruin follows. It has been demonstrated that only a small amount is necessary to absorb the carbonic acid given off by the fish and amphibians ; consequently, if the water be daily aërated with a syringe, it will absorb an abundant supply of oxygen for the animal life, and the trouble arising from the decay of much vegetable matter will be lessened or altogether avoided.

AQUARIUM, To mend broken glass of an.—Fasten a strip of glass over the crack, inside the aquarium, using for a cement white shellac dissolved in $\frac{1}{8}$ its weight of Venice turpentine.

AWNINGS, WATERPROOFING.—Dip first in a solution containing 20 per cent soap, and afterwards in another solution containing the same percentage of copper. Wash afterwards.

BENZOLE, Necessity of care in use of.—Benzole is often employed for removing grease-spots. It is highly volatile and inflammable; so that the contents of a 4-oz. phial, if overturned, will render the air of a moderate-sized room highly explosive. Never handle it near a fire or light, as the flame, igniting the vapor from an uncorked bottle, will leap over to the latter, often over a distance of several feet.

BITES, RATTLESNAKE, Remedy for.—The following is used by soldiers on the plains, and is said to be efficacious: Iodide of potassium, 4 grains; corrosive sublimate, 2 grains; bromine, 5 drachms. Ten drops, diluted with a tablespoonful or two of brandy, wine, or whisky, is the dose, to be repeated if necessary. Keep in a well-stoppered phial.

BOOT JELLY AND SHIRT COFFEE.—Some time ago, Dr. Vander Weyde, of New-York City, regaled some friends not merely with boot jelly, but with shirt coffee, and the repast was pronounced by all partakers excellent. The doctor tells us that he made the jelly by first cleaning the boot, and subsequently boiling it with soda, under a pressure of about two atmospheres. The tannic acid in the leather, combined with salt, made tannate of soda, and the gelatin rose to the top, whence it was removed and dried. From this last, with suitable flavoring material, the jelly was readily concocted. The shirt coffee, which we incidentally mentioned above, was sweetened with cuff and collar sugar, both coffee and sugar being produced in the same way. The linen (after, of course, washing) was treated with nitric acid, which, acting on the lignite contained in the fibre, produced glucose, or grape sugar. This, roasted, made an excellent imitation coffee, which an addition of unroasted glucose readily sweetened.

BOOTS, WATERPROOFING.—Use a piece of paraffine candle about the size of a nut, dissolved in lard-oil at 140° Fahr. Apply once a month.

BOOTS, To stop squeaking of.—Drive a peg into the middle of the sole.

BOOTS, Wet.—When the boots are taken off, fill them quite full with dry oats. This grain has a great fondness for damp, and will rapidly absorb the least vestige of it from the wet leather. As it quickly and completely takes up the moisture, it swells and fills the boot with a tightly-fitting last, keeping its form good, and drying the leather without hardening it. In the morning, shake out the oats and hang them in a bag near the fire to dry, ready for the next wet night; draw on the boots, and go happily and comfortably about the day's work.

BOTTLES CONTAINING RESINOUS SOLUTIONS, To clean.—Wash with caustic alkaline lyes and rinse with alcohol; if they have held essential oils, wash with sulphuric acid and rinse with water.

BOTTLES, Sealing.—Gelatine mixed with glycerine is used for this purpose. This is liquid while hot, but an elastic solid when cold.

BOTTLES, To cut in two.—Turn the bottle as evenly as possible over a low gaslight flame for about 10 minutes; then dip steadily in water, and the sudden cooling will cause a regular crack to encircle the side at the heated place, allowing the portions to be easily separated.

BOTTLES, To prevent breakage in packing.—Slip rubber rings over them.

BURNS, Remedy for.—White lead rubbed to a paste in linseed-oil. Another good remedy is as follows : Take the best white glue (extra), 15 ozs. ; break it into small pieces, add to it 2 pints cold water, and allow it to become soft. Then melt it on a water-bath, add to it 2 fluid ounces glycerine and 6 drachms carbolic acid, and continue the heat on the water-bath until a glossy, tough skin begins to form over the surface in the intervals of stirring. The mixture may be used at once, after the glue is melted and the glycerine and carbolic acid are added ; but when time allows, it is advisable to get rid of a little more of the water, until the proper point is reached. On cooling, this mixture hardens to an elastic mass, covered with a shining parchment-like skin, and may be kept for any time. When using it, it is placed for a few minutes on the water-bath until sufficiently liquid for application. (It should be quite fluid.) Should it at any time require too high a heat to become fluid, this may be corrected by adding a little water. It is applied by means of a broad brush, and forms in about two minutes a shining, smooth, flexible, and nearly transparent skin. It may be kept for any time, without spoiling, in delf or earthen dishes or pots turned upside down.

BUTTER, RANCID, To purify.—Melt in twice its weight of boiling water and shake well. Pour the melted butter into ice-water, to regain its consistence. Another : Wash in good new milk, in which the butyric acid, which causes the rancidity, is freely soluble. Wash afterward in cold spring-water. Another plan is to beat up ¼ lb. good fresh lime in a pail of water. Allow it to stand for an hour, until the impurities have settled. Then pour off the clear portion, and wash the butter in that.

CAPS, PAPER, To make.—Provide a sheet of moderately thick brown paper, size from 18 inches to 2 feet, shape as in Fig. 1. Smooth it out perfectly flat, and double over as in Fig. 2. Turn it round with the fold from you, and mark the exact middle of the piece at A, Fig. 3. Then bring down both corners, and measure off on the edge B from the point A, Fig. 3, a distance equal to ¼ the circumference of your head. Mark the point. Now turn the paper over so that the under side will be uppermost, and bend the apex of the triangle back from the point just marked, as in Fig. 4. Fold over the sides, Figs. 5 and 6, and with scissors cut off the portion, C, below the dotted line, and also the points of the two lower corners of the pieces just bent over. Next unfold the paper; spread it out flat : you will find a square marked in the middle, and creases leading therefrom to the corners of the paper. Double up the material on these creases, so as to bring up the paper as sides of a box, of which the middle square is the bottom, as in Fig. 7. Smooth the folds flat, and your work will appear as in Fig. 8. Lastly, turn

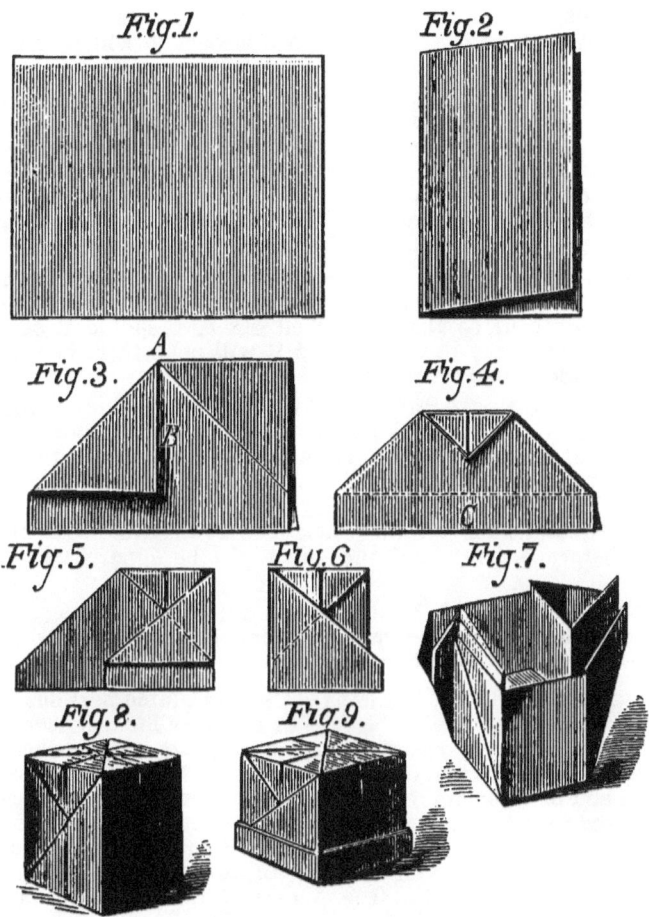

MAKING PAPER CAPS.

up the edges of the box all around twice, folding the paper on itself. Your cap is then complete, and if the measurement directed above was correctly made, it will exactly fit your head.

CALICO, To wash, without fading.—Infuse 3 gills of salt in 4 quarts of water. Put in the calico while the solution is hot, and leave until the latter is cold. It is said that in this way the colors are rendered permanent, and will not fade by subsequent washing.

CARPETS, To prevent moth in.—Wash the floor before laying with spirits of turpentine or benzine. Do not do this with a fire in the room, or with any matches or lights near.

CASKS.—Rancid butter, pork, and lard casks may be purified by burning straw or shavings in them.

CELLARS, Dry-rot in.—This, in cellar timbers, can be prevented by coating the wood with whitewash to which has been added enough copperas to give the mixture a pale-yellow hue.

CELLARS, Testing, for dampness.—Provide yourself with a thermometer, a glass tumbler filled with water, and a piece of ice ; then notice how low your thermometer, when placed in the tumbler, has to sink before any moisture begins to show itself on the outside of the vessel of cold water. The lower the temperature to which the thermometer has to sink before moisture is precipitated, the less there is of it in the moisture of the cellar.

CHAIR-BOTTOMS, To restore elasticity of cane.—Turn the chair-bottom upward, and with hot water and a sponge wash the cane ; work well, so that it is well soaked ; should it be dirty, use soap ; let it dry in the air, and it will be as tight and firm as new, provided none of the canes are broken.

CHAPPED HANDS.—Rub over with fine soap, and, while the lather is still on, scrub the hands thoroughly with about a tablespoonful of Indian meal. Rinse with tepid water, dry thoroughly, and wet again with warm water containing a quarter of a teaspoonful of pure glycerine. Dry without wiping, rubbing the hands together until all the water has evaporated. Do this at night before retiring, and the effect will be apparent by morning.

CHEST-PROTECTOR, A simple.—A folded newspaper placed over the chest inside the vest, on going out during raw spring weather, constitutes an excellent protector for the lungs.

CHICKEN FEATHERS, Utilizing.—Cut the plume portions of the feathers from the stem, by means of ordinary hand-scissors. The former are placed in quantities in a coarse bag, which, when full, is closed and subjected to a thorough kneading with the hands. At the end of five minutes, the feathers become disaggregated and felted together, forming a down perfectly homogeneous and of great lightness. It is even lighter than natural eider-down, because the latter contains the ribs of the feathers, which give extra weight. About 1.6 troy ounces of this down can be obtained from the feathers of an ordinary-sized pullet. It is suggested that, through the winter, children might collect all the feathers about a farm, and cut the ribs out as we have stated. By the spring-time, a large quantity of down would be prepared, which could be disposed of to upholsterers, or employed for domestic uses. Goose-feathers may be treated in a similar manner, and thus two thirds of the product of the bird utilized, instead of only about one fifth, as is at present the case. The chicken-down is said to form a beautiful cloth when woven. For about a square yard of the material, a pound and a half of down is required. The fabric is said to be almost indestructible, as, in place of fraying or wearing out at folds, it only seems to felt the tighter. It takes dye readily, and is thoroughly water-proof.

CHICORY, Determination of, in ground coffee.—Gently strew the powder upon the surface of cold water. Chicory, burnt sugar, etc., contain no oil, and their caramel is very quickly extracted by the water, with production of a brown color, while the particles themselves rapidly sink to the bottom of the water. On stirring the liquid, coffee becomes tolerably uniformly diffused without sensibly coloring the water, while chicory and other sweet roots quickly give a dark-brown turbid infusion. Roasted cereals do not give so distinct a color.

CHIMNEYS, BURNING, Prevention of.—The soot in the chimney can not burn, except as the fire of the stove is communicated to it through the pipe. If the pipe, therefore, be kept clean and free from soot, and the damper in the stove always closed, the chimney will never burn out. To free the pipe of soot, take the stove-handle or any convenient implement, and rap the pipe smartly on all sides from top to bottom. The soot will fall into the stove and be harmlessly consumed, or it can be removed in the usual way.

If there be a horizontal pipe, this should be taken down twice a year and thoroughly cleaned. Or, if the pipe be only a few feet in length, and the arrangements will admit of it, provide the horizontal pipe with a permanent scraper, as follows : To the end of a stout wire, a few inches longer than the pipe, attach a small segment of a disk of sheet-iron, at right angles to the wire. Remove the elbow, and thrust the scraper into the pipe. Pass the other end of the wire through a hole punched in the elbow, loop the end of the wire for a handle, and replace the elbow. After first rapping the pipe, the soot can all be drawn out and let fall into the stove. Clean the pipe thus as often as once a fortnight during cold weather.

CHIMNEYS, Smoky stove or range.—To prevent smoking, use a screen or blower of fine wire gauze, from 36 to 40 wires to the inch, immediately in front of the fire and about two inches therefrom.

CHIMNEYS, SOOTY, Cure for.—Plaster the inside with salt mortar. The proportions used are 1 peck salt added, while tempering, to 3 pecks mortar. Chimneys thus treated have remained perfectly clean for fifteen years.

CLOTHES, Fire-proof wash for.—Four parts borax and 3 parts Epsom salts, mixed with 3 or 4 parts warm water to 1 part of the combined substances, is an excellent fire-proof wash for clothes. It should be used immediately after preparation.

CHINA AND GLASS WARE, Care of.—One of the most important things is to season glass and china to sudden change of temperature, so that they will remain sound after exposure to sudden heat and cold. This is best done by placing the articles in cold water, which must gradually be brought to the boiling point, and then allowed to cool very slowly, taking several hours to do it. The commoner the materials, the more care in this respect is required. All china that has any gilding upon it may on no account be rubbed with a cloth of any kind, but merely rinsed first in hot and afterward in cold water, and then left to drain till dry. It may be rubbed with a soft wash-leather and a little dry whiting ; but this operation must not be repeated more than once a year, otherwise the gold will most certainly be rubbed off, and the china spoilt. When the plates, etc., are put away in the china closet, pieces of paper should be placed between them to prevent scratches on the glaze or painting, as the bottom of all ware has little particles of sand adhering to it, picked up from the oven wherein it was glazed. The china closet should be in a dry situation, as a damp closet will soon tarnish the gilding of the best crockery. In a common dinner-service, it is a great evil to make the plates too hot, as it invariably cracks the glaze on the surface, if not the plate itself. The fact is, when the glaze is injured, every time the "things" are washed the water gets to the interior, swells the porous clay, and makes the whole fabric rotten. In this condition they will also absorb grease; and when exposed to further heat, the grease makes the dishes brown and discolored. If an old, ill-used dish be made very hot indeed, a teaspoonful of fat will be seen to exude from the minute fissures upon its surface. These latter remarks apply more particularly to common wares.

As a rule, warm water and a soft cloth are all that is required to keep glass in good condition; but water-bottles and wine-decanters, in order to keep them bright, must be rinsed out with a little muriatic acid, which is the best substance for removing the "fur" which collects in them. This acid is far better than ashes, sand, or shot; for the ashes and sand scratch the glass, and if any shot is left in by accident the lead is poisonous.

Richly-cut glass must be cleaned and polished with a soft brush, upon which a very little fine chalk or whiting is put; by this means the lustre and brilliancy are preserved.

CIDER-STAINS.—These may be removed by lemon-juice or citric acid.

CINDERS, Sifting.—To insure servants doing this, and to prevent vegetable refuse being thrown into the ash-barrel, provide a grated cover for the latter, which secure by a padlock to prevent removal.

CISTERNS, Cleaning.—This should be done just before warm weather sets in, and should be done every year.

CLINKERS, To remove, from stoves or fire-brick.—Put in about half a peck of oyster-shells on top of a bright fire. This may need repeating.

CLOTHING, Winter.—Sufficient clothing should be worn to keep off a feeling of chilliness when about usual avocations. Less than that subjects one to an attack of dangerous pneumonia at any day or hour. More than that oppresses. Steadily aim, by all possible ways and means, to keep off a feeling of chilliness, which always indicates that a cold has been taken.

CLOTHES, Protective power of.—Clothes protect the body, by allowing through their interstices such ventilation that the nervous system may not be sensible to extremes in changes of temperature. Dr. Pettenkofer states that equal surfaces of various materials are permeated by air as follows, flannel being taken as 100: Linen of medium fineness, 58; silk, 40; buckskin, 58; tanned leather, 1; chamois leather, 51.

CLOTHES, Renovating old.—Two ozs. common tobacco boiled in 1 gallon water is used by the Chatham-street dealers for renovating old clothes. The stuff is rubbed on with a stiff brush.

The goods are nicely cleaned, and, strange to add, no tobacco smell remains.

CLOTHES, Washing compound for.—The German washerwomen use a mixture of 2 ozs. turpentine and 1 oz. spirits of ammonia well mixed together. This is put into a bucket of warm water, in which ¼ lb. soap has been dissolved. The clothes are immersed for 24 hours and then washed. The cleansing is said to be greatly quickened, and 2 or 3 rinsings in cold water remove the turpentine smell.

COCKROACHES, To drive away.—Poke-root (*phytolacca decandra*), sliced thin and laid about a house, will destroy cockroaches quicker than any other poison. It never fails. Another way of preparing the root is to boil 1 oz. in 1 pint water, until all the strength is extracted. Mix with molasses, and spread on plates in the localities infested by the insects.

CORK, To remove a, when pushed in a bottle.—When a cork gets pushed down into the neck of a bottle, insert a loop of strong twine and engage the cork in any direction most convenient. Then give a strong pull, and the cork will generally yield sufficiently to be withdrawn.

CORN, To can green.—Dissolve 2½ ozs. tartaric acid in a pint of water. Of this solution, use 1 tablespoonful to every pint of corn while the corn is at boiling heat. When opened for use, add 1 teaspoonful soda to every 3 pints of corn.

CUSHIONS, Stuffing.—Flaxseed and tallow are used in Germany as a stuffing for cushions. One part of tallow to 10 parts of flaxseed are employed, the mobility of the greased seed rendering the cushion very soft and pliable.

DISINFECTANT FOR THE BREATH, ETC.—A very weak solution of permanganate of potash is an excellent disinfectant for light purposes, such as rinsing spittoons, neutralizing the taint of diseased roots of teeth, cleansing the feet, and keeping the breath from the odor of tobacco-smoke. Permanganate is not poisonous.

ENGRAVINGS, To clean mildewed or stained.—Moisten the paper carefully, and suspend it in a large vessel partially filled with ozone. To evolve the latter, the simplest way is to clean pieces of phosphorus and place them, half covered with water, in the bottom of the jar in which the pictures are hung. On a large scale, a Ruhmkorff coil, giving a constant discharge of electricity, would be preferable.

EYE, To remove substances from the.—Take hold of the upper eyelid, near its angles, with the index-finger and thumb of each hand. Draw it gently forward, and as low down as possible over the lower eyelid, and retain it in this position for about a minute, taking care to prevent the tears from flowing out. When, at the end of this time, you allow the eyelid to resume its place, a flood of tears washes out the foreign body, which will be found adhering to, or near, the lower eyelid.

FERMENTATION OF FOOD.—This should be guarded against as the warm weather approaches. This action is always liable to cooked vegetables when set aside. Instead of warming up cold messes, it is better to scald them.

BUGS, FLEAS, ETC., To destroy.—This mixture, which has been patented in France, consists of 80 parts of bisulphide of carbon and 20 parts of essence of petroleum.

FLOORS, Cheap paint for.—This is made of 5 lbs. French ochre, ¼ lb. glue, and 1 gallon hot water. When well dried, apply one or two coats of linseed-oil.

FLOORS, Oak stain for.—An oaken color can be given to new pine floors and tables by washing them in a solution of copperas dissolved in strong lye, a pound of the former to a gallon of the latter. When dry, this should be oiled, and it will look well for a year or two ; then renew the oiling.

FLOOR WAX, Preparation of.—Heat to boiling 2 ozs. of pearl-ash, 10 ozs. of wax, and ½ pint of water. Stir frequently, until a thick fluid mass is formed from which, upon removal from the fire, no watery liquid separates out. Add boiling water cautiously, until no watery drops are distinguishable. Place on the fire again, but do not allow to boil, and add by degrees 8 or 9 pints of water, stirring constantly.

EARTHENWARE, POROUS, To clean.—This often becomes foul with organic matter when used to hold water. Use 1 oz. mu-riatic acid, rubbed on exterior and interior with a piece of flan-nel. Wash afterward with hot water.

EGGS, To preserve.—Mix together in a tub or vessel 1 bushel of quicklime, 32 ozs. of salt, 8 ozs. cream of tartar, with as much water as will reduce the composition to a sufficient consistence to float an egg. It is said that this treatment will preserve the eggs perfectly sound for two years at least.

FLOWERS, Preserving.—The flowers must be carefully sur-rounded by perfectly dry, fine sand, in such a manner that they will hold their form, the pressure of the sand upon all surfaces being alike. Any fine clean sand will answer ; it should be sifted to remove all coarse particles, and then washed in successive waters until dust and all earthy and clayey matters are washed away, and the last waters when poured off are perfectly clear. The sand is then to be dried and then placed over a fire in a proper vessel, until quite hot, hotter than the hand can bear, and when cool it will be fit to use. After heating, it should be used at once, before it can absorb moisture from the air. Good results have been obtained by taking a clean, thoroughly dry flower-pot, the hole in the bottom of which was stopped by a cork. This was filled a third full of the dry sand, the flowers set carefully in the sand, and then more sand slowly added, so as to surround and cover the flowers inside and out, and set in a warm place. At the end of 24 hours, the cork was removed from the hole in the flower-pot, and the sand allowed to run out in a small and gentle stream. The flowers were left in the pot, perfectly dry.

FLY-PAPER, ADHESIVE.—Smear paper with a mixture of mo-lasses and linseed-oil.

FRAGMENTS OF METAL, Extracting, from the flesh.—A simple and usually successful mode of extracting a needle, or any piece of steel or iron broken off in the flesh, is accomplished by the application of a simple pocket magnet. Iron filings have a way of imbedding themselves in the eye which defies almost every ordinary means for their extraction. For their removal, a small, blunt, pointed bar of steel, well magnetized, will be found excellent, and we should recommend that workmen liable to such injuries keep such an instrument about them. It would be a good plan to insert such a bar in a penknife, in a manner similar to a blade.

FRUIT, Canning.—The following table for boiling fruit in cans will doubtless prove useful. The first figure after the name of the fruit refers to time of boiling in minutes, the second to ounces of sugar to the quart : Cherries, 5, 6 ; raspberries, 6, 4 ; blackberries, 6, 6 ; gooseberries, 8, 8 ; currants, 6, 8 ; grapes, 10, 8 ; plums, 10, 8 ; peaches (whole), 15, 4 ; peaches (halves), 8, 4 ; pears (whole), 30, 8 ; crab-apples, 25, 8 ; quinces (sliced), 15, 10 ; tomatoes, 30, none ; beans and peas, 3 to 4 hours.

FURNACE HEAT, To moisten.—Dry furnace heat, productive of throat and lung diseases, may be moistened by hanging a wet towel in front of the register, the lower edge of the towel being allowed to dip in a shallow vessel of water.

FURNACE, To prevent rust in a.—Throw some quicklime loosely on a board, and place inside the furnace.

FURNITURE, Refinishing oiled or varnished.—Oiled furniture, scratched or marred, may be restored by rubbing with a woolen rag dipped in boiled linseed-oil. Varnished, by similarly rubbing with a varnish of shellac dissolved in alcohol.

FABRICS, To make uninflammable.—The lightest materials are rendered uninflammable by washing in a concentrated neutral solution of tungstate of soda, diluted with about one third of water, and then mixed with 3 per cent of phosphate of soda.

FEET, Frosted.—These can be relieved of soreness by bathing in a weak solution of alum.

FERNS, Ornaments made of.—Handsome ornaments can be made by mounting fern-leaves on glass. The leaves must first be dyed or colored. They are then arranged on the mirror according to fancy. A butterfly or two may be added. Then a sheet of clear glass of the same size is placed on top, and the two sheets secured together at the edges and placed in a frame.

FIRE-ALARM, A simple and good.—An old gun loaded with a heavy charge of powder, and hung near the rafters in a barn, or in any dangerous locality about the house, makes an excellent fire-alarm. The explosion is caused by the heat.

FIRE, Extinguishing.—A solution of pearlash in water, thrown upon a fire, extinguishes it instantly ; the proportion is 4 ozs., dissolved in hot water, and then poured into a bucket of common water.

FIRES, Kerosene.—Never try to extinguish a kerosene fire with water. Smother the flames with blankets or rugs.

FIRE, Precautions in case of.—Keep all doors and windows of the structure closed until the firemen come ; put a wet cloth over the mouth and get down on all fours in a smoky room ; open the upper part of the window to get the smoke out. If in a theatre, keep cool. Descend ladders with a regular step, to prevent vibration. If kerosene just purchased can be made to burn in a saucer by igniting with a match, throw it away. Put wirework over gaslights in show-windows; sprinkle sand instead of sawdust on floors of oil stores; keep shavings and kindling-wood away from steam-boilers, and greasy rags from lofts, cupboards, boxes, etc.; see that all stove-pipes enter well in the chimney, and that all lights and fires are out before retiring or leaving place of business; keep matches in metal or earthen vessels, and out of the reach of children; and provide a piece of stout rope, long enough to reach the ground, in every chamber. Neither admit any one, if the house be on fire, except police, firemen, or known neighbors; nor swing lighted gas-brackets against the wall; nor leave small children in a room where there are matches or an open fire; nor deposit ashes in a wooden box or on the floor; nor use a light in examining the gas-meter. Never leave clothes near the fireplace to dry; nor smoke or read in bed by candle or lamp light; nor put kindling-wood to dry on top of the stove; nor take a light into a closet; nor pour out liquor near an open light; nor keep burning or other inflammable fluids in a room where there is a fire; nor allow smoking about barns or warehouses.

FIRES, Usual causes of.—*Churches and lecture-rooms of all descriptions.*—Hot air, hot water and steam pipes, and furnaces and stoves. Sticking candles against coffins in vaults. Christmas and other decorations around or too near gas-fittings, fires, or lights. Sparks falling upon birds' nests in spires and belfries.

Curriers and workers in leather.—Lime slaked by rain. Sparks from foul flues and furnaces passing through opening and projecting eaves of drying-rooms. Friction of machinery in bark-mills. Timber, coals, shavings of wood, and leather too near flues. Drying stoves and furnaces. Spontaneous ignition. Smoking in bark and other rooms.

Drapers, tailors, makers up and vendors of male and female attire.—Working late, being tired and falling asleep, or becoming careless too near fires and lights. Unprotected and swinging gas-brackets. Crinolines coming in contact with fire in open fire-places. Light, pendent goods being blown, by the opening and shutting of doors or by concussions or drafts, into unprotected lights. Goods hung on lines increase the risk in various ways, such as conveying the flame from one end of a room to the other, and, when the line breaks down, making three separate fires, one at each end and one in the middle at the same time, thus originating three distinct fires for each line. Cuttings left carelessly about. Using lights while intoxicated, especially by tailors' work-people. Ironing-stoves, hot plates, smoothing-irons, etc., too near and sometimes on timber and goods. Smoking-tobacco, and matches for lighting it.

Engineering works, and workers in metal of all descriptions.—Sparks from striking hot metal, hot metal castings, etc., left too near timber. Heat from furnaces, forges, and smiths' hearths and flues. Friction of machinery. Japanners' stoves overheated or defective. Accidents with melted or hot metal. Explosions of blast-furnaces. Spontaneous ignition of oily waste, moulders' lamp, and other blacks ; sawdust or sweepings and oil ; spontaneous heating of iron turnings, etc., when mixed with water and oil.

Farming-stock, stables, hay, grain, or flour stores of all descriptions.—Stacking hay while green. Sparks from passing locomotives, etc. Sparks from steam thrashing machines. Sticking candles against walls and timber in barns and stables. Vagrants smoking in stables. Vagrants being refused alms. Fire-arms used near farming-stock, such as haystacks, etc.

Makers of gunpowder, fireworks, lucifer matches, and explosive compounds.—Overheating of drying-stoves and explosive mixtures. Dropping lucifers. Unprotected lights. Smoking. Leaving phosphorus uncovered with water. Friction and percussion from nails in boots. Sparks passing through broken windows. The sun's rays being concentrated through bull's-eyes, knots, etc., in glass. Defective casks containing gunpowder or other explosive materials. Spontaneous ignition of red fire and suchlike compositions. Carelessness in the supervision of young children employed. Shavings and chips too near fires and lights.

Gas-works.—Hot coke near timber, etc. Seeking for an escape with unprotected lights. Timber too near furnaces, retorts, etc. Lime slaked by rain. Defective fittings and appliances. Spontaneous ignition of coals.

Hat manufactories.—Boiling shellac. Hot irons left on timber and other inflammable things. Defective drying and other stoves. Smoking tobacco.

FISHING, Comfortable.—A plan practiced on the Western lakes in winter consists in having a small house, built on runners like those of a sled, in which is placed a small stove, while in the floor a small aperture is left through which to drop the lines. Holes are cut in the ice, the houses are moved over them, and the fishermen sit by a warm stove while drawing in the fish.

FISH, GOLD, Treatment of.—Seth Green says this as to the proper care and treatment of gold-fish : " Never take the fish in your hand. If the aquarium needs cleaning, make a net of mosquito-netting and take the fish out in it. There are many gold-fish killed by handling. Keep your aquarium clean, so that the water looks as clear as crystal. Watch the fish a little. and you will find out when they are all right. Feed them all they will eat and any thing they will eat—worms, meat, fish-wafer, or fish-spawn. Take great care that you take all that they do not eat out of the aquarium ; any decayed meat or vegetable in water has the same smell to fish that it has to you in air. If your gold-fish die, it is attributable, as a rule, to one of three causes—handling, starvation, or bad water."

FISHING-LINES, To water-proof.—Apply a mixture of 2 parts boiled linseed-oil and 1 part gold size ; expose to the air, and dry.

FLANNELS, To wash.—Take soft water, as warm as you can bear your hands in. Make a strong suds, well blued. In washing fine flannels, wet but one piece at a time ; soap the dirty spots and rub with the hands, as washboards full the flannels. When half clean, add three times as much blue as for cotton clothes. Use plenty of soap. When clean, have ready a rinse of the same temperature as the suds, rinse well, wring tight, shake briskly for a few minutes, hang out in a gentle breeze. When nearly dry, roll smooth and tight for an hour or two. Press with a moderately hot iron. If embroidered, press on the wrong side. Flannels washed in this way will look white and clean when worn out, and the quality will look better than when new.

GARBAGE, To dispose of.—When not fed to pigs, the best way to get rid of kitchen refuse is to burn it in the range or stove.

GAS ESCAPING, To detect.—To find the leak, first see that no burners have been left accidentally turned on. This is often the case where the cock has no stop, and is caused by the cock being partially turned around again so as to open the vent. Imperfect stop-cocks for this reason are dangerous, and should be promptly repaired. Try all the joints of the gas-fittings, by bringing a lighted match near them, to ignite the escaping gas if any there be. In case it is found by the sense of smell that the gas is escaping either within the floor or walls, do not on any account apply a match near a crevice. Turn off the gas at the metre, and send for a gas-fitter at once. In ordinary leaks, the burner or joint should be unscrewed, and white lead or common bar-soap rubbed in the threads before screwing home again.

GAS-LIGHT, Average prices of, in the United States.—Maine, $3.87. New-Hampshire, $3.96. Vermont, $4.80. Massachusetts, $3.86. Rhode-Island, $3.35. Connecticut, $4.03. New-York, $3.88. New-Jersey, $3.80. Pennsylvania, $3.46. Delaware, $3.95. Maryland, $3.59. Dist. of Columbia, $3.16. Virginia, $3.89. West-Virginia, $3.11. North-Carolina, $6.67. South-Carolina, $3.80. Georgia, $5.07. Florida, $8.00. Alabama, $4.83. Mississippi, $5.25. Michigan, $3.43. Wisconsin, $3.87. Ohio, $3.32. Indiana, $3.54. Illinois, $3.87. Kentucky, $3.92. Tennessee, $4.06. Minnesota, $4.31. Iowa, $4.52. Missouri, $3.95. Arkansas, $5.00. Louisiana, $4.50. Texas, $5.75. Kansas, $4.55. Colorado, $5.00. Utah, $4.00. California, $6.11.

Total average net price of gas in the United States, $4.32½.

GILT FRAMES, To restore.—Rub with a sponge moistened in urine or turpentine.

GLASS, To break in any required form.—Make a small notch, by means of a file, on the edge of a piece of glass; then make the end of a tobacco-pipe, or a rod of iron about the same size, red-hot in the fire, apply the hot iron to the notch, and draw it slowly along the surface of the glass in any direction you please; a crack will be made in the glass and will follow the direction of the iron. Round glass bottles and flasks may be cut in the middle by wrapping round them a worsted thread dipped in spirits of turpentine, and setting it on fire when fastened on the glass.

GLASS JARS, To cut.—Fill the jar with lard-oil to where you want to cut the jar; then heat an iron rod or bar to red heat, immerse it in the oil; the unequal expansion will check the jar all round at the surface of the oil, and you can lift off the top part.

GLASS, To cut without a diamond.—Hold it level under water, and, with a pair of scissors, clip it away by small bits from the edges.

GREASE-SPOTS ON CLOTHING, To remove.—In using benzole or turpentine, people make the mistake of wetting the cloth with the turpentine and then rubbing it with a sponge or piece of cloth. The only way to radically remove grease-spots is to place soft blotting-paper beneath and on top of the grease-spot, which spot has first been thoroughly saturated with the benzole, and then well pressed. The fat gets now dissolved and absorbed by the paper, and entirely removed from the clothing.

HAMS, Pickle for curing.—An excellent, well-recommended pickle for curing hams is made of 1½ lbs. of salt, ¼ lb. of sugar, ¼ oz. of saltpetre, and ½ oz. of potash. Boil all together till the dirt from the sugar has risen to the top and is skimmed. Pour it over the meat, and leave the latter in the solution for 4 or 5 weeks.

HEARTHS, SOAPSTONE, To wash.—Use pure water, and then rub with powdered marble or soapstone put on with a piece of the same stone.

HEARTHS, To clean gray marble.—Rub with linseed-oil, and no spots will show.

ICE-WATER, To preserve.—Make a hat-shaped cover of two thicknesses of paper, with cotton batting ¼ inch thick between. Place over the entire pitcher

INCUBATOR, A cheap.—One of the easiest constructed forms of incubator for the artificial hatching of eggs consists simply of a cask well buried in a manure-heap. In the bottom of the cask place one or two sieves to hold the eggs, and make a door in the side for the removal of chickens, etc. A pane of glass may also be inserted either in the door or at any convenient point for viewing the interior. In the head, which should be removable, make an opening provided with a sliding cover, for regulating the size of the aperture, as may be necessary. Form a bed of fresh manure about 1 ft. thick (after bedding) and 6 ft. square. On this set the cask, and pack more manure around the latter until flush with the top. Now take off the head or cover and place a thermometer on one of the sieves. Replace the cover. The natural heat of the manure will warm the interior of the cask. When the temperature reaches 104° (seen by the thermometer), place the eggs on the sieves. The hatching process then begins, and lasts the usual time. Care should be taken to turn the eggs over once a day, and to allow them to cool slightly, thus imitating the natural habit of the hen when she leaves her nest in search of food. The temperature of the interior is kept uniform at 104° by removing manure from the side of the cask to lessen the heat, or by substituting manure fresh from the stables in place of the older material, in order to increase the warmth.

After the chickens have emerged from the shell, the interior of the cask should be carefully cleaned, and an artificial "mother" placed inside. This last consists of a loosely-fitting disk of wood, covered on its under side with sheepskin or a piece of buffalo-robe. Under it the chickens nestle. It may be supported from the head by a piece of cord, or by a rod held in clamps, so that its distance from the bottom of the cask may be adjusted to suit. The warmth necessary for the young chickens is maintained by the manure, so that the latter answers both for this purpose and for the hatching. The slide mentioned above, as located in the head of the cask, is intended for ventilating the interior.

This plan is now in practical operation on one of the largest poultry farms in the country and is evidently more simple than any other involving the use of special apparatus and gas or lamps for heating. A cylindrical vessel must be used—never a square one, since the chickens, in the latter case, will crowd into corners and smother each other. The number of eggs hatched depends upon the size of the cask or the number of casks used. As many as one thousand eggs have been thus incubated at a time. Any farmer having a manure-heap, however small, can easily test the plan, if only with a dozen eggs. The matter requiring the greatest care is to keep the temperature in the cask uniform, and to have the manure sheltered from rain, which would cool it.

INK, INDELIBLE, To remove.—If the ink is a nitrate of silver preparation, it may be taken out of the fabric (1) by washing the latter in a solution of hyposulphite of soda, or (2) by moistening it with a solution of bichloride of copper, and then washing it with liquid ammonia.

INK-STAINS, To remove, from mahogany.—Put a few drops of spirits of nitre in a teaspoonful of water, touch the spot with a feather dipped in the mixture, and, on the ink disappearing, rub it over immediately with a rag wetted in cold water, or there will be a white mark which will not easily be effaced.

INK-STAINS, To remove.—Wash carefully with pure water, and apply oxalic acid ; and, if the latter changes the dye to a red tinge, restore the color by ammonia.

INSECT BITES.—A good remedy is borax, 1 oz., dissolved in 1 pint water previously boiled and allowed to cool.

KEYS, Fitting.—When it is not convenient to take a lock apart to fit a new key, the key-blank should be smoked over a candle, inserted in the keyhole, and pressed firmly against the opposing wards of the lock. The indentations in the smoked portion made by the wards will show where to file.

LAMP-BURNERS, To fasten kerosene.—Plaster of Paris mixed with resin soap is a good cement for this purpose.

LEAD-COLIC, Preventives of.—If working in lead, wash the hands several times a day in a strong decoction of oak-bark. Keep the hair short, and (if a painter) wear a clean cloth cap. The clothes should be frequently washed, and the hands also, especially before touching food. Before eating, the mouth should be rinsed with cold water. A weak oak-bark decoction should be used as a wash several times a week. The body should be sponged night and morning with cold or tepid water, and the hair thoroughly washed every evening after work. The food should contain a large proportion of fatty substances, and milk should be taken in large quantities.

LEAF AND FLOWER IMPRESSIONS, To make.—Take a small quantity of printer's ink, thinly put it on glass, evenly distributed. The end of the index-finger will serve as the printer's ball, to cover one side of the leaf uniformly ; then lay it to the exact place where you wish the print to be ; lay over it a piece of thin, soft paper large enough to cover it ; then, without moving the leaf, press all parts of it with the end of the thumb firmly, and you will have a perfect impression, that no engraver can excel ; and by adjusting the leaves at the proper points, accurate prints can be taken, and, aided with the brush or pen, the stem and whole plant can be shown. Excellent specimens of impressions of barks of trees can be made by slicing the bark ; and with a little care, the stems can also be taken, as well as flowers. When colored with the aniline colors, they are very like colored engravings.

LEATHER, To water-proof.—Saturate with castor-oil. This is excellent for winter boots.

LIFE PRESERVER, A simple.—It is not generally known that, when a person falls into the water, a common felt hat can be made use of as a life-preserver. By placing the hat upon the water, rim down, with the arm around it pressing it slightly to the breast, it will bear a man up for hours.

LINEN, To bleach.—Javelle water, used for turning white the dirtiest linen and removing stains, is composed of bicarbonate of soda, 4 lbs. ; chloride of lime, 1 lb. Put the soda into a kettle over the fire, add 1 gallon of boiling water, let it boil from ten to fifteen minutes, then stir in the chloride of lime, avoiding lumps. Use when cool. This is good for removing fruit-stains from white underwear.

MARBLE, To clean.—Common soda, 2 parts ; pumice-stone (pulverized), 1 ; finely powdered chalk, 1. Sift through a fine sieve, and mix with water. Rub all over the marble until the stains are removed. Then wash the stone with soap and water. Marble that is yellow with age, or covered with green fungoid patches, may be rendered white by first washing it with a solution of permanganate of potash of moderate strength, and while yet moist with this solution, rubbing with a cloth saturated with oxalic acid. As soon as the portion of the stone operated upon becomes white, it should be thoroughly washed with pure water to remove all traces of the acid.

MATCH-SCRATCHERS.—The best are pieces of shark-skin, or squares of fine wire gauze.

MICE, To kill.—Sprinkle some grain near the holes, and throw near by a few bits of cotton saturated in chloroform. This has been tested, and mice have been found dead, two or three at a time, lying with their noses near the cotton.

MILDEW, To remove.—Make a very weak solution of chloride of lime in water (about a heaping teaspoonful to a quart of water), strain it carefully, and dip the spot on the garment into it ; and if the mildew does not disappear immediately, lay it in the sun for a few minutes, or dip it again into the lime-water. The work is effectually and speedily done, and the chloride of lime neither rots the cloth nor removes delicate colors, when sufficiently diluted and the articles rinsed afterward in clear water.

Moss Ornaments.—A beautiful orrament for the sitting-room can be made by covering a common glass tumbler with moss, the latter fastened in place by sewing-cotton wound around. Then glue dried moss upon a saucer, into which set the tumbler, filling it and the remaining space in the saucer with loose earth from the woods. Plant the former with a variety of ferns, and the latter with wood-violets. On the edge of the grass also plant some of the nameless little evergreen vine, which bears red (scarlet) berries, and whose dark, glossy, ivy-like foliage will trail over the fresh blue and white of the violets with beautiful effect. Another good plan is to fill a rather deep plate with some of the nameless but beautiful silvery and light green and delicate pink mosses, which are met with in profusion in all the swamps and marshes. This can be kept fresh and beautiful as long as it is not neglected to water it profusely once a day. It must, of course, be placed in the shade, or the moss will blanch and die. In the centre of this, a clump of large azure violets should be placed, adding some curious lichens and pretty fungous growth from the barks of forest-trees, and a few cones, shells, and pebbles.

Mosquitoes, To drive off.—Rub the skin with essence of pennyroyal, or with a little coal-oil on a bit of cotton. The smell of the oil disappears in a few minutes.

Mustard Poultice, To make a.—In making a mustard plaster use no water, but mix the mustard with white of egg, and the result will be a plaster which will draw perfectly, but will not produce a blister, no matter how long it is allowed to remain.

Mucilage, Pocket.—Boil 1 lb. best white glue, and strain very clear; boil also 4 oz. isinglass, and mix the two together; place them in a water-bath (glue-kettle) with ½ lb. white sugar, and evaporate till the liquid is quite thick, when it is to be poured into moulds, dried, and cut into pieces of convenient size. This immediately dissolves in water, and fastens paper very firmly.

Newspaper Binder, Temporary.—Take two pieces of light wire, strong enough to reach across the paper once, and three or four pieces of stout thread. Place one wire under the paper as far from the edge as you choose to bind it. Put the threads around the lower wire up through the paper, and tie them over the other wire on top. Temporary covers of stiff pasteboard may be added, having holes for the reception of the thread, the wires being placed on the outside of the cover. The successive papers are, of course, to be threaded, one by one, by means of an awl or coarse needle.

Oil-Cloths, Cleaning.—These should not be washed with soap. A coat of good copal varnish at long intervals improves them. Oil-cloths should never be scrubbed. Wipe with a wet cloth, after brushing with a soft floor-brush.

Oiled Floors.—The scrapings from these should immediately be placed in the open air. They are liable to spontaneous combustion.

OIL-PAINTINGS, To restore old.—Take the painting out of the frame, lay it on a table, face up, and keep a wet cloth on it for two or three days, changing or cleaning the cloth as often as it becomes soiled. When the painting is clean, wash it with a sponge or brush dipped in nut-oil. This is much better than varnishing

PAINTER'S COLIC.—(1.) One drachm of sulphuric acid in 10 pints of table or spruce beer, or mild ale. Shake well, and allow it to stand for a few hours. Take a tumblerful twice or three times daily. (2.) Make a beer of molasses, 14 lbs.; bruised ginger, ½ lb.; coriander-seed, ½ oz.; capsicum and cloves, ¼ oz. each; water, 12½ galls.; yeast, 1 pint. Put the yeast in last, and let it ferment. When the fermentation has nearly ceased, add sulphuric acid, 1¼ ozs., mixed with 12 ozs. water, and 1½ ozs. bicarbonate of soda dissolved in water. It will be fit to drink in three or four days.

PAINT, To clean.—Dip a flannel rag into warm water, and wring it out nearly dry. Take up on the rag as much whiting as will adhere, and rub this on the paint until the dirt or grease disappears. Wash the part well with clean water, and rub dry with soft chamois-skin.

PAINT, To remove, from clothes.—Chloroform will remove paint from a garment or elsewhere, when benzole or bisulphide of carbon fails.

PAINT, To remove old.—Slake 3 lbs. of stone quicklime in water, and add 1 lb. American pearlash, making the whole into the consistence of paint. Lay over the old work with a brush, and let it remain for from 12 to 14 hours, when the paint is easily scraped off.

PAPER COMFORTERS.—Two thicknesses of paper are better than a pair of blankets, and much lighter for those who dislike heavy bedclothes. A spread made of double layers of paper tacked together, between a covering of chintz or calico, is really a desirable household article. Soft paper is the best, but newspapers will answer.

PAPERING WALLS.—Papering and painting are best done in cold weather, especially the latter, for the wood absorbs the oil of paint much more than in warm weather; while in cold weather the oil hardens on the outside, making a coat which will protect the wood instead of soaking into it. Never paper a wall over old paper and paste. Always scrape down thoroughly. Old paper can be got off by damping with saleratus and water. Then go over all the cracks of the wall with plaster of Paris, and, finally, put on a wash of a weak solution of carbolic acid. The best paste is made out of rye-flour, with 2 ozs. glue dissolved in 1 qt. paste; ½ oz. powdered borax improves the mixture.

PASTE, A superior flour.—Thoroughly mix good clean flour with cold water to a paste, then add boiling water, stirring up well until it is of a consistence capable of being easily spread with a brush. Add to this a little brown sugar, a little corrosive sublimate, and about half a dozen drops of oil of lavender; and keep, if convenient, two days before using.

PETROLEUM BARRELS.—These should not be used to store food or drink in. They are poisonous even after being cleaned.

PLANT-CASE, A housetop or window.—A fernery or plant-case might be arranged to run the whole length of the front windows of a story, and be heated by a small boiler placed behind a fire-

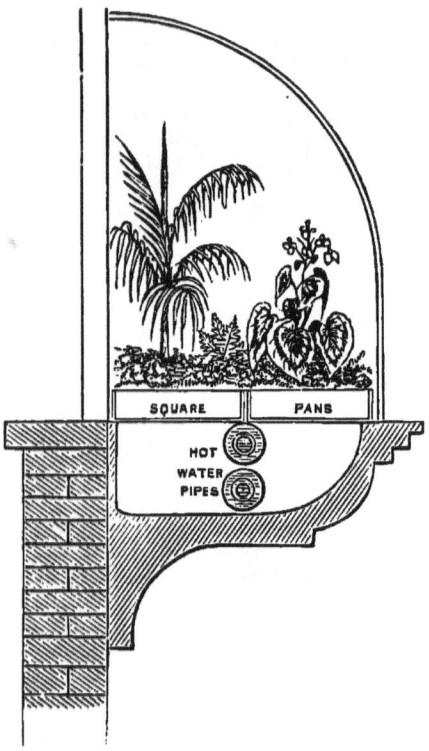

SECTION OF PLANT-CASE.

place.   From this a 2-inch flow and return pipe is taken through the case, so as to heat it when required. The space around the pipes can be filled with bark, or water if desirable, so as to produce a moist and genial bottom heat. The ferns, mosses, and other decorative plants, are arranged in flat square pans of zinc or earthenware, as shown in our sectional sketch, and the effect of the whole, especially when seen from within, is very effective, and affords relief to the eye, which might otherwise look out on a dismal prospect of blackened roofs and soot-begrimed chimney-pots.

PLANTS, WINDOW, Care of.— Plants kept in the windows should be turned every morning, or the light, striking on one side only, will draw the plant to that side, so that all its branches and leaves will turn toward the window. The water in the saucers should never be applied to the plants. In cutting slips of any plant, always choose the youngest branches ; and cut off the slip at the junction of a joint or leaf, since the roots shoot more readily from such joints. If you follow these directions, and put sufficient sulphate of ammonia to just taint the water applied to your plants, you may cultivate with success almost any plant, even though you are an entire novice.

PLANT-CASES, WARDIAN, Management of.—The following principles are those upon which a fern-case should be constructed : 1. Have no apparatus or arrangement for drainage. 2. Make your case as air-tight as possible, allowing for no ventilation.

Ferns require, for their growth, shade and moisture ; upon the former, in a great degree, depends the latter. A northern or eastern aspect, where the morning sun reaches the case, we think is best. As regards moisture, we have the principle of self-support in an air-tight case ; for if you allow the sun to reach the case for an hour or so in the morning, you will find that the moisture needful for the growth of your ferns is extracted from the earth ; and when evening comes, this same moisture will condense and fall. Each day, this process of extraction and condensation takes place, and your plants flourish under a necessary and sufficient moisture. Now, this being the kind of air we want, we must not, of course, ventilate our case, and allow it to escape, otherwise the dry air of our rooms would enter, and the watering of the case become a necessity. This at once upsets all the benefits derived from these cases. The temperature, also, must be much more even in an air-tight case than in a ventilated one, where the constant opening and shutting of doors and windows would affect it. If we have no watering to do, we have no water to run off, and consequently require no drainage in the bottom of our case. Now, in this air-tight principle, we get at the secret.

In stocking Wardian cases, the amateur will find that almost all ferns and mosses will do well in this case. There are few of our greenhouse ferns that will not do well under this treatment ; the gold and silver ferns are perhaps the exception ; they do not always attain their full size and beauty in a Wardian case, but the adiantums, pteris, polypodiums, blechnums, and others do well.

In planting a case, do not place the plants too near, nor use too many of a large size, but put in a few plants and of a moderate size. Water well after setting the plants out, and shade the case for a day or two ; then give it the morning sun each day for an hour or two, and your ferns will soon start. Nothing can be more interesting than to watch them—the frond pushes its head above the earth, the heat and moisture of the case have their effect, and it gradually rises and uncurls till it reaches its height, then it expands into the most beautiful and graceful of shapes ; then what can exceed in delicacy and freshness this newly-born part ? The lycopodiums grow finely, and spread very rapidly in the case ; small pieces introduced at regular intervals in the case will, in a marvelously short time, double their original size ; and if the pendent roots of the creeping species are pressed well on to the surface of the earth, the spaces between the plants and ferns will soon be filled up, and a rich and delicate carpet be produced

over the whole case.   For climbers, nothing can give more satis-
faction than *ficus stipulata*, which can be obtained at all green-
houses.   The roots of this plant, which strike out at every joint,

THE PRINCESS OF WALES CASE.

have an adhesive power, and will attach themselves firmly to the
glass in the case, which renders the growth more rapid and regular.   It is a very interesting plant to watch ; the roots adhering
to the glass allow a free use of the microscope, and the growth
and circulation can be studied to great advantage from the outside of the case.

THE SYDENHAM CASE.

As to soil, the best mixture for the growth of ferns and
lycopodiums is the following : Leaf-mould, 2 parts ; fresh
sand, 1 part ; gravel, about the size of a pea, 1 part ; and
stable manure, chopped very fine, 1 part. Ferns which grow naturally in dry places can be arranged on rock-work in the centre

of the case, if it is large enough to admit of it, and those requiring more moisture should be placed nearer the sides of the case, and they will get more moisture from the glass, where it deposits in great quantities. The spores of ferns can be sown on the surface of the earth in the Wardian case, and a constant supply of young plants can in this way be obtained, thus enabling the student to watch them in every stage of development.

It happens that not unfrequently the larvæ of insects are introduced in the earth into the case, and hatch out under the influence of the heat. To provide against this, it will be found useful and interesting to put in a small-sized toad, and insects will disappear very soon, and give no further trouble. Toads will live through the winter perfectly well in this way, and their habits can be studied ; some may become aware, by trying this experiment, that the toad, although not one of the handsomest of our reptiles, is not the least interesting.

PLANTS, Potting.—Those who find their efforts to raise house-plants frustrated by worms may be able to win success by boiling the earth before setting the plants. Use little water, and allow it to simmer away after a few minutes of hard boil.

POLISH, FURNITURE.—Shave very fine white wax, 3 ozs., cas-tile-soap, 1 oz. ; put the wax in 1 gill turpentine, and let it stand 24 hours. Boil the soap in 1 gill water, and add to wax and turpentine.

POTATOES, SARATOGA, Fried.—The following is all there is of the cook's secret for producing those world-renowned potatoes served at Moon's Lake House, Saratoga Springs, every summer : Peel good-sized potatoes, and slice them as evenly as possible ; drop them into ice-water. Have a kettle of lard, as for fried cakes, and very hot. Put a few at a time into a towel, shake them about to dry them, and then drop into the hot lard. Stir them occasionally ; and when of a light brown, take them out with a skimmer. If properly done, they will not be at all greasy, but crisp without, and mealy within.

POTATOES, Frozen.—These can be cured by soaking in water 3 days before cooking.

RUST-SPOTS, To remove from cloth.—Wet the spots of iron-rust on muslin or white dress-goods thoroughly with lemon-juice, then lay in the hot sun to dry. Repeat the same if the color is not removed by one application. When dry, rinse in clear, cold water. Lemon-juice can not be used on colored goods, as it will take out printed colors as well as stains. It will remove all kinds of stains from white goods.

RING, To remove, when tight on the finger.—In case a finger-ring becomes too tight to pass the joint of the finger, the finger should first be held in cold water to reduce any swelling or inflammation. Then wrap a rag soaked in hot water around the ring to expand the metal, and lastly soap the finger. A needle threaded with strong silk can then be passed between the ring and finger, and a person holding the two ends and pulling the silk, while sliding it around the periphery of the ring, will readily remove the latter. Another method is to pass a piece of sewing-silk un-

der the ring, and wind the thread in pretty close spirals and closely around the finger to the end—that below the ring—and begin
unwinding.

RICE, To boil.—The way they boil rice in India is as follows :
Into a saucepan of 2 quarts water, when boiling, throw a tablespoonful of salt ; then put in 1 pint rice, previously well washed
in cold water. Let it boil 20 minutes, throw out in a colander,
drain, and put back in the saucepan, which should be stood near
the fire for several minutes.

RAIN-WATER, To preserve sweet.—A drachm of pounded alum
to a gallon of water is sufficient. After 24 hours, the water will
be cleansed. All wooden vessels to hold water should be charred
inside. If a mixture in the proportion of ¼ lb. of lime, made into
a paste, and added to a spoonful of powdered alum, be put into 200
gallons of water, it will soften the water, and precipitate vegetable and other matter.

RATS, Bait for.—Put a drop of rhodium oil on a bit of cheese or
meat. These animals detest chloride of lime and coal-tar.

RATS, To catch.—Cover a common barrel with stiff paper, tying
the edge around the barrel. Place a board so that the rats can
have easy access to the top. Sprinkle cheese or other bait on the
paper, and allow the rats to eat there unmolested for several days.
Then place in the bottom of the barrel a stone 6 or 7 inches high,
and pour in water until all the stone is covered, except for a space
about big enough for one rat to crawl upon. Now replace the
paper, first cutting a cross in the middle. The first rat that
climbs on the barrel-top goes through into the water, and
climbs on the stone. The paper comes back to its original position, and the second rat follows the first. Then begins a fight
for the possession of the dry place on the stone, the noise of which
attracts the others, who share the same fate.

RAZORS, Paper for sharpening.—By merely wiping the razor on
the paper to remove the lather after shaving, a keen edge is maintained without further trouble. The razor must be well sharpened at the outset. First, procure oxide of iron (by the addition of
carbonate of soda to a solution of persulphate of iron), well wash
the precipitate, and finally leave it of the consistence of cream.
Spread this over soft paper very thinly with a soft brush. Cut
the paper in pieces two inches square, dry, and it is ready for
use.

RAZOR-STROP, To make a.—Select a piece of satin, maple, or
rose wood, 12 inches long, 1¼ inches wide, and ⅜ inch thick ; allow
3½ inches for length of handle. Half an inch from where the
handle begins, notch out the thickness of the leather so as to
make it flush toward the end. Taper also the thickness of the
leather ; this precaution prevents the case from tearing up the
leather in putting the strop in. Then round the wood very
slightly, just enough (say $\frac{1}{12}$ of an inch) to keep from cutting by
the razor in stropping and turning over the same. Now select a
proper-sized piece of fine French bookbinder's calfskin, cover with
good wheat or rye paste, then lay the edge in the notch, and secure it in place with a small vise, proceed to rub it down firmly

and as solid as possible with a tooth-brush handle (always at hand, or should be), and, after the whole is thoroughly dry, trim it neatly and make the case.

SLEEPLESSNESS, Cure for.—Mr. Frank Buckland says : " If I am much pressed with work, and feel I shall not sleep, I eat two or three small onions, and the effect is magical. Onions are also excellent things to eat when much exposed to intense cold. In salmon-fishing, common raw onions enable men to bear the ice and cold of the semi-frozen water much better than spirits, beer, etc. If a person can not sleep, it is because the blood is in his brain, not in his stomach ; the remedy, therefore, is obvious : call the blood down from the brain to the stomach. This is to be done by eating a biscuit, a hard-boiled egg, a bit of bread and cheese, or something. Follow this up with a glass of wine or milk, or even water, and you will fall asleep."

SMOKED MEAT, To preserve.—The keeping qualities of smoked meat do not depend upon the amount of smoking, but upon the uniform and proper drying of the meat. It is of considerable advantage also to roll the meat on its removal from the salt, before smoking, in sawdust or bran. By this means the crust formed in smoking will not be so thick ; and if moisture condenses upon the meat it remains in the bran, the brown coloring matter of the smoke not penetrating. The best place to keep the meat is in a smoke-house, in which it remains dry without drying out entirely, as it does when hung in a chimney.

SPATTER-WORK PICTURES.—These are delicate designs in white appearing upon a softly-shaded ground. Procure a sheet of fine uncalendered drawing-paper, and arrange thereon a bouquet of pressed leaves, trailing vines, letters, or any design which it is desired to have appear in white. Fasten the articles by pins stuck into the smooth surface, which should be underneath the paper. Then slightly wet the bristles of a tooth or other brush in rubbed Indian ink, or in common black writing-ink, and draw them across a stick in such a manner that the bristles will be bent and then quickly released. This will cause a fine spatter of ink upon the paper. Continue the spattering over all the leaves, pins, and paper, allowing the centre of the pattern to receive the most ink, the edges shading off. When done, remove the design, and the forms will be found reproduced with accuracy on the tinted ground. With a rustic wooden frame, this forms a very cheap and pretty ornament.

SPOONS, To remove stains on, caused by boiled eggs.—Rub with common salt.

SPONGES, Cleaning.—A gelatinous substance frequently forms in sponges after prolonged use in water. A weak solution of permanganate of potassa will remove it. The brown stain caused by the chemical can be got rid of by soaking in very dilute muriatic acid. An old and dirty sponge may be cleaned by first soaking it for some hours in a solution of permanganate of potassa, then squeezing it, and putting it into a weak solution of hydrochloric acid, 1 part acid to 10 parts water.

SPONGES, To bleach.—Wash first in weak muriatic acid, then in cold water ; soak in weak sulphuric acid, wash in water again, and finally rinse in rose-water.

STAINS OF ACID FRUIT, To remove, from the hands.—Wash the hands in clear water, wipe them lightly, and while they are yet moist, strike a match and shut your hands around it so as to catch the smoke, and the stain will disappear.

STARCH, To prevent souring when boiled.—Add a little sulphate of copper.

STONE, To remove moss from.—This is useful for the green mould which forms on marble and brown-stone steps. Apply a solution of 75 grains of carbolic acid to 1 quart of water.

STOPPERS, GLASS, To remove.—To move a tight glass stopper, hold the neck of the bottle to a flame, or take two turns of a string and seesaw it. ·The heat engendered expands the neck of the bottle before the expansion reaches the stopper.

STOVE-HOLES IN WALLS.—See that these openings for the pipes are protected by good tin covers after the stoves are taken down. Do not stuff rags in.

STRAW MATTING, Washing,—Use a cloth dampened in salt water. Indian meal sprinkled over it and thoroughly swept out will also cleanse it finely.

STYPTIC PAPER, for stopping the bleeding of small wounds.— Mix gum benzoin (best quality), 1 lb.; rock alum, 1 lb.; water, 4½ gals. Boil in a tin vessel for 4 hours, replacing the water lost by evaporation. Saturate paper with the solution, dry carefully, and brush over with a concentrated solution of perchloride of iron. Keep in a water-proof and air-tight case.

SAFE, Home-made fire-proof.—The best is a hole in the ground well lined with brick and cement.

SHIRT-BOSOMS, Glossing,—Take 2 ozs. of fine white gum-arabic powder, put it in a pitcher, and pour on a pint or more of water, and then, having covered it, let it stand all night. In the morning, pour it carefully from the dregs into a clean bottle, cork, and keep it for use. A teaspoonful of gum-water stirred in a pint of starch, made in the usual way, will give to lawns, white or printed, a look of newness when nothing else can restore them, after they have been washed.

SHOES, Bronzing.—Black shoes may be bronzed by a strong solution of aniline red in alcohol.

SHOES, Black varnish for.—Take 10 parts, by weight, of shellac, and 5 of turpentine. Dissolve in 40 parts alcohol, in which fluid should be previously dissolved 1 part extract of logwood, with some neutral chromate of potassa and sulphate of indigo. This varnish is to be kept in well-stoppered bottles.

SIDEWALKS, Slippery.—Put on hard sand instead of ashes.

SILK, Washing.—The way to wash silk is to spread it smoothly upon a clean board, rub white soap upon it, and brush it with a clean hand-brush.

SILVER, To clean.—A strong solution of hyposulphite of soda is useful for this purpose.

SILK, ETC., To clean.—A teaspoonful of powdered borax dissolved in 1 qt. tepid water is good for cleaning old black dresses of silk, cashmere, or alpaca.

SILVER-PLATE, To keep bright.—Warm the articles, and coat carefully over with thin collodion diluted with alcohol, using a wide soft brush for the purpose.

SINK-SPOUTS, To thaw frozen.—Place one end of a piece of lead pipe against the ice to be thawed, and then through a funnel in the upper end pour boiling water. Keep the pipe constantly against the ice, and a foot or more per minute can be penetrated. In order to thaw out water-pipes that become frozen and are inaccessible, the plan used by New-York plumbers is to surround small india-rubber tubing with coiled wire so as to stiffen it and admit of its being inserted far into the pipe. Through the tube a current of steam from a small boiler over a charcoal furnace is allowed to pass. This acts very quickly, except when the pipe takes a very irregular course, in which case there is no remedy except to dig down into the earth or break out walls until the pipe can be reached, and thawed by the direct application of heat.

SOAP, Adulteration of, by starch.—This is detected by dissolving the soap in alcohol, which leaves the starch behind.

SOAP, GALL, To make.—Gall soap, excellent for washing silks and ribbons, may be made by heating 1 lb. cocoanut-oil to 60° Fahr., into which ½ lb. caustic soda is gradually stirred. To this ½ lb. Venice turpentine, previously warmed in another vessel, is added. The kettle is allowed to stand for four hours, subject to a gentle heat, after which the fire is increased until the contents are perfectly clear. One pound ox-gall, followed by 2 lbs. castile-soap, is then mixed in, and the whole allowed to cool, when it may be cut into cakes.

SOAP, HARD, To prevent crumbling.—Dip the bars in a mixture of resin-soap, beef-tallow, and wax.

SOAP, Home-made.—Soap-making is not an easy process; sometimes the ashes are poor, or the right proportions of lye and grease are not used; at other times the soap appears to be good when put up, but changes entirely after standing a few days. The last trouble usually arises from getting the soap too strong and diluting with water. If very strong, it will be thin and dark; and by adding cold water and thoroughly stirring, the color is changed many shades lighter and the mass thickened, giving it the appearance of a No. 1 article, while in reality it is very poor. Hickory-ashes are the best for soap-making, but those from sound beach, maple, or almost any kind of hard wood except oak, will answer well. A common barrel, set upon an inclined platform, makes a very good leach; but one made of boards set in a trough in V-shape is better, for the strength of the ashes is better obtained, and it may be taken to pieces when not in use, and put away. First, in the bottom of the leach, put a few sticks; over them spread a piece of carpet or woolen cloth, which is much better than straw; put on a few inches of ashes, and then from 4 to 8 qts. lime; fill with moistened ashes, and pack well down. Pack the finest in the centre. It is difficult to obtain the full strength of ashes in a barrel without removing them after a day's leaching, and mixing them up and replacing. The top should be first thrown off and new ashes added to make up the proper quantity. Use boiling water for second leaching. Take about 4 gallons lye, and boil up thoroughly with 12 lbs. clear grease, then add the lye as it is obtained, keeping a slow fire and stirring often until you have a barrel of soap. After boiling the grease and 4 gallons lye together, it may be put in a barrel and the rest of the lye added there. This will form good soap if frequently stirred ; but the heating process is the best, when weather and time will permit.

TATTOO-MARKS ON THE SKIN, To remove.—Blister the part with a plaster a little larger than the mark ; then keep the place open for a week with an ointment ; finally, dress it to get well. As the new skin grows, the tattoo-marks will disappear.

TAR-SPOTS, To remove.—Butter will remove tar-spots. Soap and water will afterward take out the grease-stain.

TEA-KETTLE, To prevent rust forming inside a.—Keep an oyster-shell in the bottom of the kettle ; and when water is wanted, pour off without agitating the vessel. Be careful also not to let the water stand in the vessel when not in use.

TEETH, Extracting, Simple method of, for children.—The operation consists in simply slipping a rubber ring over the tooth and forcing it gently under the edge of the gum. The patient is then dismissed, and told not to remove the appendage, which in a few days loosens the tooth, and causes it to fall out.

TIN, Scouring.—Kerosene and powdered lime, whiting, or wood-ashes, will scour tins with the least labor.

TOOTHACHE.—Saturate a bit of cotton wool in a strong solution of ammonia, and apply it immediately to the affected tooth.

TUBS AND PAILS, to prevent shrinking of.—Saturate with glycerine.

VEGETABLES, To wash.—Vegetables should never be washed until immediately before prepared for the table. Lettuce is made almost worthless in flavor by dipping it in water some hours before it is served. Potatoes suffer even more than other vegetables through the washing process. They should not be put in water till just ready for boiling.

VENTILATION OF SLEEPING-ROOMS, Simple plan for.—A piece of wood 3 in. high, and exactly as long as the breadth of the window, is to be prepared. Let the sash be now raised, the slip of wood placed on the sill, and the sash drawn closely upon it. If the slip has been well fitted, there will be no draft in consequence of this displacement of the sash at its lower part ; but the top of the lower sash will overlap the bottom of the upper one, and between the two bars perpendicular currents of air, not felt as draft, will enter and leave the room.

VINEGAR, To make, from molasses.—Vinegar may be made by mixing 16 parts pure water, 1 part syrup of molasses, and 1 part baker's yeast at a temperature of about 80° Fahr., and keeping the compound in a warm atmosphere from ten to thirty days. A little old vinegar, added on the second or third day, will aid the process.

VINEGAR, Raspberry.—Pour over 1 lb. bruised berries, 1 qt. of the best cider vinegar ; next day, strain the liquor on 1 lb. of fresh ripe raspberries, bruise them also, and on the following day do the same. Do not squeeze the fruit, only drain the liquor thoroughly. Put the juice into a stone jar, and add sugar in proportion of 1 lb. to a pint. When the sugar is melted, place the jars in a saucepan of water, which heat ; skim the liquor, and after it has simmered for a few minutes, remove from the fire, cover, and bottle.

WASHING-BLUE.—Twenty lbs. white potato starch, 20 lbs. wheat starch, 20 lbs. Prussian blue, 2 lbs. indigo carmine, and 2 lbs. finely-ground gum-arabic are mixed in a trough, with the gradual addition of sufficient water to form a half-fluid, homogeneous mass, which is then poured out on a board with strips tacked to the edges. It is then allowed to dry in a heated room until it does not run together again when cut. It is next cut, by a suitable cutter, into little cubes, and allowed to dry perfectly. They are finished by being placed in a revolving drum, with a suitable quantity of dry and finely pulverized Paris blue, until they have a handsome appearance. The cost is about 12 cents per pound.

WASHING COLORED FABRICS.—Before washing almost any colored fabrics, soak them in water, to each gallon of which a spoonful of ox-gall has been added. A teacupful of lye in a pail of water is said to improve the color of black goods. A strong tea of common hay will improve the color of French linens. Vinegar in the rinsing water, for pink and green, will brighten those colors ; and soda answers the same end for both purple and blue.

WALL-PAPER, Removing stains on.—Stains on wall-paper can be cut out with a sharp penknife, and a piece of paper so nicely inserted that no one can see the patch.

WARTS, Cure for.—Rubbing warts, night and morning, with a moistened piece of muriate of ammonia is said to cause their disappearance without pain or a scar resulting.

WATER-CLOSETS, Ventilating pipes for.—Extend pipes from water-closet traps, or one (larger) from the main waste-pipe, into the nearest chimneys. The pestilent gases will thus be carried off instead of being allowed to escape into the house.

WATER-PIPES, To prevent freezing of.—Have a cock in the cellar by which the water can be turned off from the house.

WATER, To purify, from smoke.—Enough permanganate of potassa is added to give the faintest possible tinge to the water. After standing 24 hours, the impurities will all be precipitated.

WOUNDS, CUT.—A wound made by a knife or other sharp instrument is best healed by bringing the edges together and putting on a bandage which will not exclude the air. Nature will work the cure, if the person be healthy, much better than any salve or ointment.

WATER-LILIES, To raise.—Water-lilies may be raised about one's house by the following method : Sink in the ground the half of an old cask, and cover the bottom with peat and swamp mud, and then fill with water. Dig the lily roots early in the spring, and place them in the earth at the bottom of the tub.

WINDOWS, Washing.—In washing windows, a narrow-bladed wooden knife, sharply pointed, will take out the dust that hardens in the corners of the sash. Dry whiting will polish the glass, which should first be washed with weak black tea mixed with a little alcohol. Save the tea-leaves for the purpose.

WINE, Preservation of, by heating.—Wine may be kept without altering in quality for an indefinite period of time, in all climates, after having been first submitted to the action of artificial heat. The temperature to which it must be raised is from 131° to 140° Fahr. If the wine does not contain naturally more than 10 or 12 per cent of alcohol, it is best to add 1½ per cent more before the shipping of it. The wine is to be heated by steam and artificially cooled.

YEAST, COMPRESSED.—Previously malted barley and rye are ground up and mixed, next put into water at a temperature of 65° to 75° ; after a few hours the saccharine liquid is decanted from the dregs, and the clear liquid brought into a state of fermentation by the aid of some yeast. The fermentation becomes very strong ; and by the force of the carbonic acid which is evolved, the yeast globules are carried to the surface of the liquid, and, forming a thick scum, are removed by a skimmer, then placed on cloth filters, drained, washed with a little distilled water, and next pressed into any desired shape by means of hydraulic pressure, and covered with a strong and well-woven canvas. It keeps from 8 to 14 days, according to the season, and is excellent.

YEAST FOR HOT CLIMATES.—Boil 2 ozs. best hops in 4 qts. water for ½ hour ; strain and cool to new-milk warmth. Put in ½ lb. sugar, 1 tablespoonful of salt ; beat up 1 lb. of the best flour with some of the liquor, and mix all well together. Let it stand for 3 days, and on the third day add 3 lbs. mashed and boiled potatoes. On the next day, strain, and it is ready for use. This will keep for 2 or 3 months in a moderately cool place. The yeast is very strong ; half the usual quantity necessary for baking is sufficient.

YEAST, VIENNA.—Vienna bread and Vienna beer are said to be the best in the world. Both owe their superiority to the yeast used, which is prepared in the following manner : Indian corn, barley, and rye (all sprouting) are powdered and mixed, and then macerated in water at a temperature of from 149° to 167° Fahr. Saccharification takes place in a few hours, when the liquor is racked off and allowed to clear, and fermentation is set up by the help of a minute quantity of any ordinary yeast. Carbonic acid is disengaged during the process with so much rapidity that the globules of yeast are thrown up by the gas, and remain floating on the surface, where they form a thick scum. The latter is carefully removed, and constitutes the best and purest yeast, which, when drained and compressed in a hydraulic press, can be kept from 8 to 15 days, according to the season.

RICHARD H. BUEL,

Mechanical Engineer

80 BROADWAY, NEW-YORK,

OFFERS HIS SERVICES TO

THOSE WHO BUY MACHINERY,

To enable them to avoid costly mistakes;

THOSE WHO SELL MACHINERY,

To enable them by tests to prove the correctness of their guarantee;

THOSE WHO USE MACHINERY,

1st. To enable them to discover defects and mismanagement;

2d. To enable them to intrust the charge of the machinery to suitable persons;

THOSE WHO HIRE POWER,

AND

THOSE WHO LET POWER,

That they may have the amount accurately determined;

THOSE WHO INVENT MACHINERY

AND

THOSE WHO FURNISH CAPITAL TO INVENTORS,

That they may have the merits of the inventions carefully investigated;

THOSE WHO NEED THE ASSISTANCE OF A RELIABLE AND COMPETENT ENGINEER,

In expert cases of any kind.

SHELDON'S AUXILIARY
CAR SEAT.

A new Folding Stool, of strong and simple construction, easily attached to the regular seats of street cars. Adds fifty per cent to the seating capacity of the car, and prevents the passage being crowded by passengers standing.

When in use, as shown in engraving, does not interfere with people already seated, and when out of use can be folded entirely out of the way. Compact, durable, and cheap. Equally well adapted to omnibuses, or for any locality where temporary seats are needed. Patented Aug. 3d, 1875.

Address the inventor,

CEVEDRA B. SHELDON,
No. 7 State Street, New-York City.

www.ingramcontent.com/pod-product-compliance
Lightning Source LLC
Chambersburg PA
CBHW031425020726
47499CB00005B/1602